RELATIVITY AND COMMON SENSE

RELATIVITY AND COMMON SENSE

A New Approach to Einstein

HERMANN BONDI

Dover Publications, Inc.
New York

Published in Canada by General Publishing Company, Ltd., 30 Lesmill Road, Don Mills, Toronto, Ontario.

This Dover edition, first published in 1980, is an unabridged and corrected republication of the work originally published in 1964 by Doubleday & Company, Inc., New York.

International Standard Book Number: 0-486-24021-5
Library of Congress Catalog Card Number: 80-65687

Manufactured in the United States of America
Dover Publications, Inc.
180 Varick Street
New York, N.Y. 10014

PREFACE

This book differs radically from all previous attempts to explain Relativity to the lay audience.

Where previous writers have tried to develop Relativity *in opposition to* the ideas of Isaac Newton, Professor Bondi derives Relativity *from* Newtonian ideas. He pictures Relativity as being neither revolutionary nor destructive of classical dynamics but rather as being an organic growth, inevitable when man began to deal with velocities approaching the speed of light.

Readers whose mathematical backgrounds are limited should have no trouble following Professor Bondi's derivation of the mathematics necessary to understanding of the elementary aspects of Relativity. For fifty years the practice has been to begin with the Lorentz Transformation, which involves systems of coordinates moving relative to each other, and then to use the Transformation to establish the concepts and characteristic effects of Special Relativity. Professor Bondi has reversed the process. He first establishes these concepts and effects and then shows how they lead, by simple algebra, to the Lorentz Transformation. Beginners thus will enjoy distinct advantages from Professor Bondi's approach to Relativity: He makes use of an understanding of Newtonian ideas to develop concepts of Relativity, and he uses relativistic concepts to derive a mathematical treatment, a logical progress in which one step leads to the next and the reader can advance with confidence.

Like its predecessor, *The Universe at Large* (1960), this book had its beginning in articles written for *The Illustrated London News,* but for the present publication the author has revised and extended the original material.

Relativity and gravitational theory are Professor Bondi's fields of professional concentration, but he is best known among lay science enthusiasts and students as one of the three principal originators of the so-called "Steady-State" theory of cosmology, Fred Hoyle and Thomas Gold being the other two. A Viennese by birth, Professor Bondi is a product of Trinity College, Cambridge, England, and is Professor of Applied Mathematics at King's College, University of London. He is a strong believer in the duty of scientists to tell society what science is about, and his occasional appearances on BBC educational programs have made him, if not a star, at least a figure of considerably more than passing interest to the English television audience.

Professor Bondi is a Fellow of the Cambridge Philosophical Society, a Fellow of the Royal Astronomical Society, and a Fellow of the Royal Society.

John H. Durston

CONTENTS

RELATIVITY AND COMMON SENSE

CHAPTER I

"ON THE SHOULDERS OF GIANTS"

When the Theory of Relativity first came out, and for many years afterward, it was looked on as something revolutionary. Attention was focused on the most extraordinary aspects of the theory. With the passage of time, though, the sensational aspects of Albert Einstein's work have ceased to cause wonderment, at least among scientists, and now one begins to see the theory not as a revolution, but as a natural consequence and outgrowth of all the work that has been going on in physics since the days of Isaac Newton and Galileo. Although the theory changed some important notions very much and quite unexpectedly, we now can see that perhaps the notions that suffered such changes were the less important ones and that the basic ideas that were fully maintained were the more important ones.

The approach that will be followed in this book may therefore be called the traditionist's approach to relativity. Since it is the tradition of physical science that enters, we have to consider many of the basic ideas of the older physics on which the Theory of Relativity is built. Science is a continuing process. Newton said beautifully of his own work, "If I have seen further than others it is because I stood on the shoulders of giants." In a book like this we have to visualize at every stage what each of the giants from Newton to Einstein thought and what the evidence for his views was. If we seem a little long in coming to relativity and, particularly, in coming to anything sensational, then it is only because of our delight in the older, as well as the newer, parts of physics, and because of the need to climb a staircase step by step.

The Concept of Force

One of the most difficult issues in science is to decide when a particular phenomenon is worth investigating. It has been said that a fool can ask more questions than a wise man can answer, but in science the problem is far more often that a wiser man is needed to ask the right question than to answer it. One of the oldest problems is the question of motion. This has puzzled people for many centuries. Why do things move as they do? What makes them move? We seem to have a natural feeling that inanimate matter left to itself will come to rest. A bouncing ball will bounce less high each time until finally it comes to rest. A rolling cart, at least on flat ground, will gradually slow down and stop. Even on the smoothest materials, such as ice, a stone sliding along will come to rest eventually. At the beginning of every motion there seems to stand a living body like the player throwing his ball. However true this observation may be on the small scale, on the larger scale it seems to break down. Thus the wind blows and, unless we invoke a god of the winds, there is no animate matter pushing it. The tides come and go and ocean currents flow and, most mysterious of all, the Moon and the planets carry on their movements across the skies. Which is the more complicated phenomenon—the Moon's continuing to circle the Earth or the ball's ceasing to bounce and coming to rest?

Since the ball is familiar and the motion of the Moon

less immediately apparent, for a long time people jumped to the conclusion that it was the motion of the ball that was the simpler, and therefore that the coming to rest was natural, that there was something about the state of rest that attracted objects to it, while a special explanation was required for the continuing motion of the planets, like Kepler's idea that angels were pushing them along their orbits. It needed the genius of Newton to see that things were the other way round. There is nothing peculiar about the state of rest: there is only a series of very complicated phenomena that we call friction preponderating in our neighborhood. In the skies we see the simpler phenomena in which friction does not matter, and celestial bodies move naturally. Before Newton the permanence of their motion was thought to require explanation–an explanation that might be thought of as force. It was only Newton who saw that the question was wrongly put. There was nothing to be explained about velocity; what did require explanation were the changes of velocity–accelerations.

The Evaluation of Acceleration

The idea of acceleration is quite basic. It is a measure of the rate of change of velocity. Velocity changes not only if the speed increases or diminishes, but also if its direction alters. There is no acceleration only if a body is moving in a straight line with constant velocity, and acceleration is the measure of the deviation of its motion from this standard. Nowadays, with smooth transport, we are in a better position to realize that velocity does not matter. Pouring out a cup of tea in the dining room at home is an operation requiring a certain minimum of skill. Pouring out a cup of tea in a jet plane flying smoothly at 600 miles per hour is precisely the same operation requiring precisely the same kind of skill. The fact that in the one case we are moving relative to the Earth, and in the other not, is totally irrelevant. Thus Newton's first great insight, his Principle of Relativity, as we might call it, is that velocity does not matter. To put it a little more precisely, what can be done inside a box is independent of the box's velocity provided only that the velocity is constant. This is familiar from dining cars. If the train is running smoothly, pouring out a cup of tea in the dining car is just as simple as at home or in the jetliner, but if the train is braking sharply or going round a bend or being jerked by crossing points, then the operation requires very much more skill—otherwise a lot of tea is spilt.

Thus as soon as the velocity ceases to be constant, as soon as there are accelerations, new factors enter. How do we evaluate this acceleration? This calculation is most easily accomplished by representing velocity by an arrow, an arrow in the direction of the motion and of a length representing the speed of the motion. If we then compare the velocity arrows at one instant and a second later, with the foot of each arrow in the same place, the arrow that runs from the tip of the first one to the tip of the second one represents the acceleration. When a train is increasing speed, then the acceleration is in the same direction as its velocity. But a case of at least as great importance is when the train is going round a bend. Then the speed stays the same, but the arrow points at successive moments in different directions. It is readily seen that the arrow joining the tips of the two arrows a few moments of time apart is more or less at right angles to the arrows themselves. The acceleration is at right angles to the velocity.

To account for accelerations, the concept of force is introduced. Force is that which is required to produce an acceleration. Again, this is not unfamiliar. Tie a stone into a net and swing the net round you. You have to exert a force to keep the stone the same distance from you. The stone is following a circular orbit and so the acceleration is at right angles to the orbit. Hence, the acceleration is toward you, and your pull on the stone (through the medium of the net) is the force that accounts for the acceleration keeping the stone in its orbit.

Another case, on a much larger scale, concerns the motion of the Earth round the Sun (Fig. 1). As long

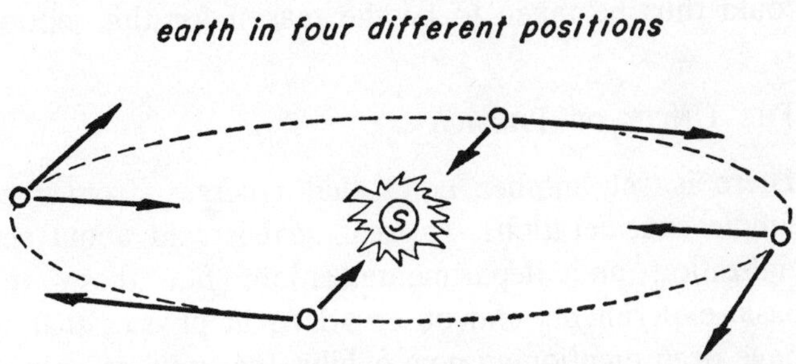

FIG. 1. *Tangential arrows point in the direction of the Earth's velocity at the four positions; the arrows at right angles pointing toward the Sun give the direction of acceleration.*

as one thinks that the *keeping in motion* requires an explanation, one naturally looks in the direction of the Earth's velocity to find the cause of the motion. But looking in that direction one never sees anything of great significance, but a different object every time one looks. If, however, one looks at right angles to the direction of the velocity, that is, in the direction of the acceleration of the Earth, then one always sees the Sun, which is clearly a body of great significance. In other words, merely changing the question from "What causes the *velocity* of the Earth?" to "What causes the *acceleration* of the Earth?" immediately leads one from chasing a hare to seeing the Sun, undoubtedly the most important object in our astronomical neighborhood. It needs, then, only a small step to say that thus, by merely changing the question, one gets from an unknown cause of the Earth's motion to the immediate idea that it is the Sun which is responsible for the orbit of the Earth. Similarly, Newton accounted for the orbit of the Moon round the Earth by showing that the acceleration of the Moon was always directed toward the Earth, which could thus be taken to be the reason for this motion.

The Unity of Physics

There is yet another point that emerges from these simple considerations. A great deal is said about specialization and departmentalization, but these very basic experiments and observations in physics that we have been mentioning now exhibit *the unity of physics*.

For purposes of textbooks and of university examinations, physics is subdivided into such subjects as dynamics, the science of force or optics, the science of light, etc., but this division is highly artificial and cannot be sustained. It is probably impossible to think of any experiment that is purely dynamical or purely optical. Some combination is always involved. So when Newton said that dynamics was fundamentally concerned with acceleration, and the acceleration of the Earth was found to point toward the Sun, such a junction of dynamics and optics had been achieved. The acceleration, the basic dynamical property of the orbit of the Earth, is in the direction in which we *see* the Sun, and to see an object is the simplest and most important optical observation. Thus the very way in which it is found that acceleration is so important in dynamics depends on an optical observation—the direction in which the light from the Sun arrives. Keeping this basic lesson of the unity of physics in mind, we shall easily be able to avoid the pitfalls that misled physics in the late nineteenth century, and to follow necessarily the track of Einstein's Theory of Relativity.

CHAPTER II

MOMENTUM

In the preceding chapter attention was given to Newton's identification of the direction of force with the direction of acceleration, and to the fact that absence of force implies absence of acceleration. The question that was not then discussed was *how much* acceleration follows from a given amount of force. The clue to the answer to this question is the concept of mass, or, a little more highbrow, the concept of momentum. As for mass, we are all familiar with the fact that the same spring extended the same way will move some objects much less rapidly than others, and we call the objects that move sluggishly massive, while the ones that move rapidly are called light. If one wanted to pursue the argument in detail, one would say that the same force, such as a spring extended a certain way, will produce accelerations of different particles in certain ratios, and this ratio is, by experiment, found to be independent of the actual extension of the spring, though, of course, the individual accelerations do depend on it. Thus we introduce the notion of mass with the idea that the acceleration produced by a given force is inversely proportional to the mass of the object on which it acts—that is, the greater the mass the smaller the acceleration. In daily life we do not usually measure mass but weight. This is what scales are for. In fact, mass and weight are exceedingly closely related, so closely that ordinarily one can use one to stand in for the other. But from the point of view of the physicists weight is a rather more complicated concept than mass, since it depends on a local circumstance, the strength of the Earth's gravitational field.

To come now to the notion of momentum, which is such a very useful notion in physics, we simply multiply mass by velocity. The masses of objects are usually constant (though not when, for example, the motion of a growing raindrop through a cloud is considered, or the trajectory of a rocket that is firing out gases at the back). The rate of change of momentum equals the force and it is this law which, in fact, generalizes to the more complicated cases just referred to. It is a remarkably useful law, because we can apply it to a whole system. The question arising here is a very deep one which goes through the whole of science. In science one is always concerned to put the best face on one's ignorance. One is never in total command of the facts, and a scientist who waits until he knows everything before he says anything is like the man who will not make a decision until he has all the facts. One never has all the facts, the scientist's knowledge is always very limited, and he has to make the best with what he has got. Thus it could be argued that it is absurd to try to evaluate the gravitational field of the Earth as it affects the Moon's orbit without knowing every detail of the internal constitution of the Earth. But this is not so. Fortunately, a very great deal can be said about the Moon's orbit and even about the orbit of a sputnik, without knowing much about the internal constitution of our Earth.

The Motion of a System of Bodies

Momentum is what the mathematician calls an additive quantity. The momentum of a large body is the sum of the momenta of all its individual particles. The individual particles may have the most complicated forces acting between them, but the momentum of the whole body depends only on the forces acting on it from outside. And it is often extremely useful to be able to say something about the motion of a whole system of bodies without knowing how every individual particle moves. An example drawn from everyday life may help. Consider a baby in a pram on smooth, level ground. At first the baby is asleep and the whole system, pram and baby, is at rest. Then the baby wakes up and begins to kick. What happens now? Can the baby propel the pram along by its kicks? See Fig. 2.

If we consider the whole system of pram and baby (and suppose the baby to be safely strapped in so that the two will not separate), then the momentum of this system can change only if external forces act. To move it horizontally there must thus be horizontal external forces. The only way in which these can arise is through the friction against the ground. For our purposes we may suppose, without changing anything essential in the picture, that if the brake is off (so that the pram can roll smoothly), then there is no friction against the ground, whereas if the brake is on, so that the wheels turn only with difficulty, then there is such friction. In the first case, with the pram unbraked, it follows that the momentum of the whole system must always vanish. Therefore the center of mass, as we call it, of the whole system, having originally been at rest, must remain at rest. Motions internal to the system are possible, how-

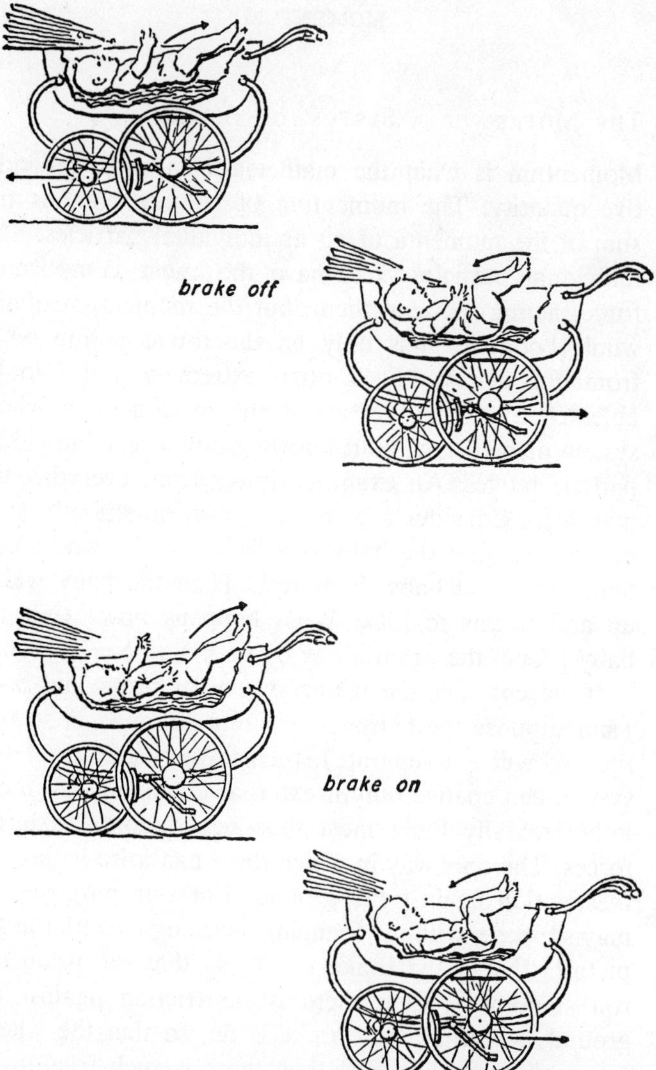

FIG. 2. *In the first two drawings, with the carriage brake off, movements of the baby's legs cause equal and opposite motions of carriage. In the last two drawings, with brake on, resulting motions are no longer equal and opposite.*

ever. Thus if the baby kicks out its feet the pram moves a little in the opposite direction by recoil, but as soon as the baby pulls its feet up the pram will move back by precisely the same amount. What happens when the brakes are on? Of course, if the brakes are perfect and the ground is very rough, then no motion is possible and we need not think about the situation. But suppose that the pram stands on fairly smooth ground so that some motion is possible. Then it may well occur that as the baby kicks its feet out this increases its weight, because it pushes its body downward, and therefore there is much more resistance to the motion of the pram and no motion will result. But in the opposite case, when the baby pulls its legs in, downward, its weight is diminished, and so the friction on the ground is less and now the pram may move a little. All the motion of the pram is now in the same direction. If the baby keeps on kicking for a long time there may therefore be a substantial motion of the pram if the brake is on, whereas with the brake off there can be no major motion, although the pram may move forward and backward a little bit each time the baby kicks. The result is rather the opposite to what one would expect, but on suitable ground and in suitable conditions the results stated here are indeed verified by observation. Of course, if the ground is not strictly level or if there is a wind blowing the pram along, it would be exceedingly dangerous to take the brake off, but, say, indoors on a smooth level floor the situation will be exactly as represented here. With the brake off the pram will move more with each of the baby's kicks, but the forward and backward movements will cancel exactly, and by and large there will be no motion. On the other hand, with the brake on, with each individual kick there will be far less motion than before, but these movements may add to something very substantial in the course of time. The remarkable thing is that we can make a very definite statement about the behavior of the pram on smooth level ground in the absence of friction. One might have thought that in order to predict the motion of the pram one would have to know how the baby kicks and would have to consult a child psychologist to discover exactly how the kicking goes on, etc. But this is not necessary. When there is no frictional force acting, we can say something about the motion of the pram without knowing what goes on inside it. It is this fact that makes the law of conservation of momentum so extremely useful. It is a law that enables us to state something about the over-all behavior of systems that we do not understand in detail. A minor point that emerges is the tremendous importance of systematic movements. It is the fact that the movements, however small, may be additive in the case of the braked pram that makes all the difference. The much larger movements of the unbraked pram add up to zero.

The Momentum of an Airplane

The law of conservation of momentum applies to many fields and is one of the most important laws of physics that we have. Thus when we think of a propeller-driven plane, then the forward momentum of the plane can be increased only by supplying backward momentum to the air. What the propeller does is to push the air backward. This involves, as a kind of recoil action, a forward push for the plane. In this sense the difference between a propeller-driven and a jet plane is purely a technological one. For some purposes it is more advantageous to use a propeller outside the engine to produce the backward flow of air; for other purposes it is more advantageous to use machinery inside the engine to produce a powerful rearward flow of air and exhaust gases. In either case, as must be true from the law of conservation of momentum, the forward push on the plane must be balanced by a similar backward motion given to other materials, such that the momentum imparted to the one is equal and opposite to the momentum imparted to the other. The rearward motion always involves both the exhaust gases of the engine and the surrounding air, but the relative importance of these two, understandably, depends on how much air there is. The higher the plane flies the thinner the atmosphere, the more important the rearward motion of the exhaust as compared to that of the surrounding air. If then, one goes right out into space where there is virtually no medium, then the forward push of a missile must be due to the rearward momentum of the exhaust gases alone. Thus spaceships or very high-flying missiles use their own exhaust as the only possible way of producing rearward momentum to balance their own forward momentum. This is the principle of a rocket, which, of course, can be applied at ground level, too, as we all find out on the Fourth of July in the United States or on Guy Fawkes' Day in England.

The Irrelevance of Velocity

In the atmosphere any increase in speed brings its own difficulties with it. There is a rest state that is distinguished from the others; the rest state of the atmosphere surrounding the plane. The faster the plane, the greater its friction against the surrounding atmosphere, the harder the engine has to work to keep the plane going at the requisite speed. But out in space the situation is quite different. It needs no more fuel to increase the speed of a spaceship from 1000 miles an hour to 2000 miles an hour than it needs to increase the speed of a spaceship of the same mass from 100,000 miles an hour to 101,000 miles an hour. By the Newtonian principle of relativity, velocity does not matter for these dynamical phenomena. A spaceship once at a speed of 10,000 miles an hour will glide along at constant speed just as happily as a spaceship at 1000 miles an hour will, and it needs just as much effort to raise the speed of the one as to raise the speed of the other. The notion that we have a real state of rest is one that is entirely due to our own surroundings, in which the atmosphere and the ground beneath us always give us an idea of rest. Once we are out of these local circumstances and have shed our parochial prejudices, every speed is just as good as any other speed, as far as this dynamical work goes. This is an immediate consequence of Newton's laws and a vital principle of this branch of physics.

CHAPTER III

ROTATION

In the previous chapters the irrelevance of velocity has been discussed. Velocity, however, is only one of the possible aspects of the motion of a simple rigid body. In addition to a motion of translation the body may possess a motion of rotation. One might at first sight suppose that just as there was no state of translatory velocity distinguishable from all others, there might also be no such distinguished state of rotary motion. This is not so. We all know that there is a state of no rotation in which a body can be without any internal strains but that, as soon as rotation sets in, the body tries to distend, and strains arise to keep the parts farthest from the axis from flying off. From what has been said before it is not difficult to see how this arises. There is no force only if a particle continues with the same velocity, both in direction and in magnitude (speed). However, a particle on a rotating body will change the direction of its velocity in the course of its motion around the center of the body. Thus there will have to be a force holding the particle in, a centripetal force, to counter the fictitious centrifugal pull of the particle. The centripetal force can vanish only if there is no rotation whatever. Hence the very law that shows that there is no preferred state of rest for a motion of translation implies that there is a state of rest as far as rotation is concerned. Not having velocity does not mean anything unless the standard of rest is specified. Non-rotating, however, means something perfectly definite.

To measure the speed of rotation we can either speak of the total time of revolution or we can speak of a subdivision. We may discuss, for example, how many seconds a body takes to turn through 1 degree, or, as the mathematician usually does, we speak about the time the body takes to turn through an angle of just over 57 degrees ($180°/\pi$ to be precise), an angle which is called one radian. The reciprocal of this time is called the angular velocity. The higher the angular velocity the faster a body is turning. Engineers work in terms of the number of revolutions per minute (r.p.m.), and when one talks about the speed of engines this is the terminology usually employed. Whichever way we count it, angular velocity is the correct measure of rotation.

Measurement of the Earth's Rotation

A body whose rotation is of particular importance to us is the Earth. How do we measure the rotation of the Earth? Perhaps the best-known method is the use of Foucault's pendulum. To see how this works we imagine an ordinary pendulum freely suspended above the North Pole and allowed to swing there. Then the pendulum moves freely under the influence of gravitation. If originally set up to swing in a straight line, it will go on doing so. The rotation of the Earth is quite irrelevant to it. The Earth will rotate beneath the pendulum. Thus to an observer on the Earth the plane of the motion of the pendulum will appear to be rotating in just the opposite way to the direction in which the Earth is rotating. If the pendulum were fixed somewhere other than at the Pole, the situation would be rather different because some of the rotation would be ineffective, as it would try to pull the pendulum out of the vertical. It is easily seen that, looking downward on the North Pole, the plane of the Foucault pendulum appears to be rotating clockwise, whereas looking down at the South Pole, at a pendulum similarly fixed there, the rotation will appear to be anticlockwise. It does not need any great mathematical insight to appreciate that in between the situation is intermediate. Thus on the Equator the pendulum would not vary its plane at all relative to the Earth; in northern latitudes it would rotate clockwise, but more slowly than at the Pole, diminishing its speed of rotation from the Pole to zero in low latitudes, and the opposite would occur in the Southern Hemisphere. Thus the Foucault pendulum is a means of measuring the rotation of the Earth.

Another way of measuring the rotation of the Earth is to look at the fixed stars. The Earth is also rotating relative to them, so that the celestial sphere seems to be revolving round the Earth. The remarkable fact is that the rotation of the Earth relative to a Foucault pendulum on the Pole is, as near as one can tell, the same as the rotation of the Earth relative to the fixed stars. This raises the question first asked by Bishop George Berkeley[1] in the eighteenth century and then put more precisely by Ernst Mach[2] toward the end of the last

[1] George, Earl of Berkeley (1685–1753) was a clergyman, philosopher, and Fellow of the Royal Society, author of *A Treatise concerning the Principles of Human Knowledge.* His concept of time as relative conflicted with Newton's picture of the universe. It took mathematicians more than a century to overcome the technical objections he raised against Newton's statement of the calculus. Berkeley was a brilliant writer and had one of the most acute minds in the history of philosophy.

[2] Ernst Mach (1838–1916) was an Austrian philosopher and physicist of great influence in modern thought. Einstein repeat-

century, and again by Einstein in the early years of this century. What is the connection between the two rates of rotation? Roughly speaking, it appears as though the distant matter of the universe determined what state of rotation was to be called no rotation in our neighborhood. The most precise measurements of what this state is probably follow from the motion of the bodies of the Solar System and appear to agree extremely well with the state of no rotation found by observing distant matter. How it is that the distant masses fix the state of no rotation is not wholly clear, although Einstein's General Theory of Relativity goes a long way toward accounting for this very mysterious fact.

The Coriolis Effect

While it might seem that the dynamic effects of the rotation of the Earth are a matter of not very much importance for our daily lives, though relatively easily ascertained by looking at the stars or at a Foucault pendulum or a gyroscope, yet it is true to say that our whole lives probably are shaped by them. The influence comes from the effect of the rotation on the motion of the atmosphere–i.e., the winds. The effect is perhaps most easily imagined on a gramophone turntable (Fig. 3). Suppose a particle is gently moved toward the axle on such a turntable, having previously been at rest relative to the turntable. It will then carry with it its forward velocity from the place it originated and, as this was farther from the axle than its later positions, it will seem to be going forward in

---

edly acknowledged his debt to Mach's analysis of Newtonian mechanics as a philosophical groundwork to formulation of Relativity. Mach numbers, the familiar measure of speed in the jet- and missile-age, are named for him.

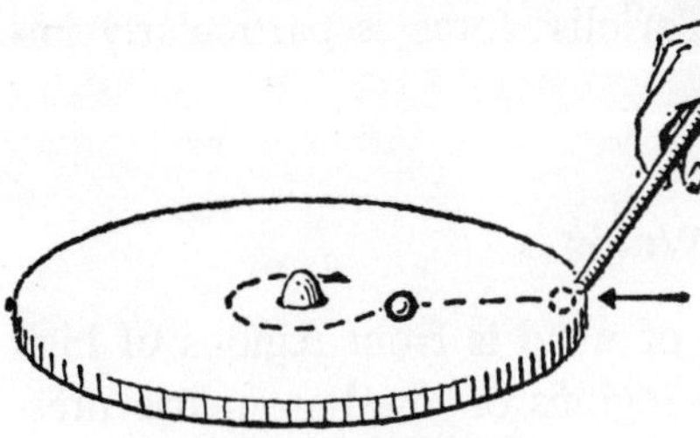

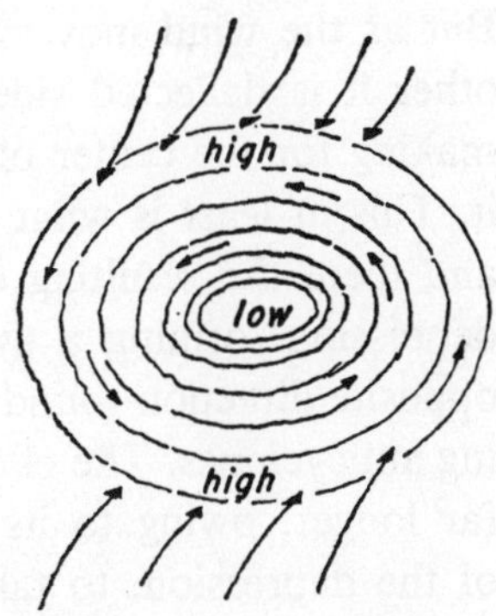

FIG. 3. *As in the experiment with the ball on the turntable, winds are deflected as they blow from one place to another owing to the rotation of our planet. Thus the winds are made to circulate instead of going straight from a high-pressure to a low-pressure region.*

the direction of rotation, relative to its new neighbors. Similarly, a particle moving in a circle round the axle relative to the turntable in a direction that enhances the rotation will thereby go faster round the axis and so will require a larger force than previously to keep it on its circular track. If no such larger force is applied it will push outward relative to its neighbors. Thus the effect of the motion of the turntable will be that any particle moving on the turntable will be deflected in a direction at right angles to that in which it first wished to go. This so-called Coriolis[3] force is particularly important for the winds.

The Movement of Winds

Basically the direction of wind is from regions of high barometric pressure to regions of low barometric pressure in order to equalize the pressures between the two. But as the wind moves from one pressure area to the other it is deflected sideways so much that instead of making for the center of the depression it circles round it. This at least is what happens in moderate latitudes, and there the resulting effect is that winds go round a depression forming a cyclone, whereas they go in the opposite direction round regions of high pressure, forming anticyclones. The effect of this is that the wind takes far longer, owing to its many orbits round the center of the depression, to take air toward the low pressure region than it would otherwise do, and hence regions of high and low pressure have such long lives. Because of this persistence of the pressure pattern, the low pressure regions can drift far inland and bring rain to regions distant from the ocean. It is due to this drift, essentially, that the continents in moderate and fairly high latitudes are fertile and not deserts. Similarly the trade winds, which also result from the rotation of the Earth, though in a slightly different way, bring moisture to the continental regions in the vicinity of the Equator. Thus the whole pattern of life and agriculture on the surface of the land is based on the fact that the Earth rotates. It may be worth mentioning that the effect of

[3] Gaspard Gustave de Coriolis (1792–1843) was a French engineer and scientist. The use of Work to mean force multiplied by distance originated with him.

the rotation of the Earth in deflecting winds brings about a situation that is quite different from what one might expect. So strong is the effect, that instead of flowing directly from high pressure to low pressure at right angles to the lines of constant pressure, the wind follows almost exactly the lines of constant pressure, the isobars.

Angular Momentum and Angular Velocity

Just as a body in translatory motion has a tendency to persist and, in fact, will keep its velocity unless a force acts, so a rotating body will have a tendency to persist in its rotation. The measure of what it is that persists in the straight-line motion we called momentum or, to be more precise, linear momentum, and our rule was that without force there could be no change of linear momentum. It was particularly important that this rule applied to a system as a whole and that we did not have to know every detail about the system to be able to speak about the fate of its momentum. All we had to know were the external forces. The measure of the tendency of a body to keep rotating is called its angular momentum, and if there are no forces acting that tend to slow down or accelerate its rotation (the technical term for such forces is couples or moments), then the angular momentum of the body will persist. However, the angular momentum of a body is not related to its angular velocity as simply as the linear momentum is related to the linear velocity, where there was just a usually constant factor of proportionality, the mass of the body. Mass is unalterable (except by throwing things out as in the case of rocket fuel), but the link between angular momentum and angular velocity is much more complicated, and this is easily understood. If all the particles making up a body are close to its axis of rotation, then they will all move only at low speeds even when the body is being spun round the axis many times per second. If, however, the shape of the body is changed so that many of its particles are far from the axis, then at the same rotation rate as before these particles will now have to move fast. Thus if the masses are far from the axis of rotation, even a small angular velocity will produce a large angular momentum, whereas if the masses are near the axis of rotation the opposite will be the case. There is then the possibility, by shifting masses, of changing the angular velocity while keeping the angular momentum constant. This is well known in skating. If a skilled skater rotates slowly with his arms outstretched and then pulls his arms in and makes himself as narrow as possible, owing to the fact that his angular momentum must stay the same, his angular velocity will increase very much, and he will pirouette at high speed. This is quite a complicated subject, and so it comes as a slight surprise that the cat understands these matters very thoroughly indeed, at least, in an instinctive way.

Dropping a Cat

We all know that whichever way you drop a cat it will land on its feet. At first sight this seems a very remarkable phenomenon. If a cat is dropped without rotation and, therefore, without angular momentum, how can it turn over so as to arrive on its feet (Fig. 4)? It must have had an angular velocity at some stage of its fall, although its angular momentum must have been zero all the time. How could the cat have had an angular

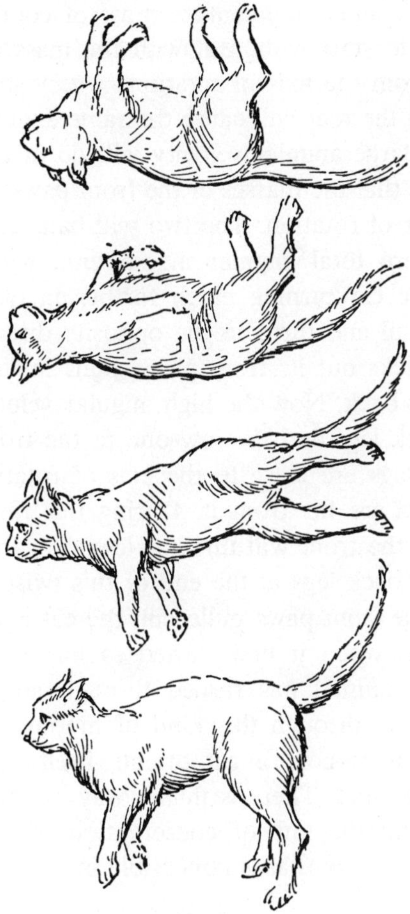

FIG. 4.

velocity without having an angular momentum? The explanation is found in the marvelous flexibility of cats.[4] Suppose first that the cat sticks out its hind legs, pulls in its front legs and head, and twists its back.

[4] The actual motion of a cat is rather complex but conforms, of course, to the principles laid down here. In order to make

The total angular momentum must of course be zero, as it was to start with. But with the mass of the hind legs far from the axis of rotation, a very small angular velocity in the rear will cause the same angular momentum as a large angular velocity will do in front, owing to the fact that the masses of the front paws are so close to the axis of rotation. The two will balance, therefore, to give zero total angular momentum, with the head end of the cat turning much farther in one direction than the tail end turns in the opposite direction. Next, the cat sticks out its front paws, pulls in its hind legs, and twists back. Now the high angular velocity belongs to the back legs and the low one to the front, because the hind legs are close to the axis of rotation and the front paws are far from it. During this time of twist, therefore, the front will turn far less than the rear end. When the back legs at the end of this twist are pushed out and the front paws pulled in, the cat is in the same position as when it first started to move, except that the whole animal has turned through an appreciable angle. Going through this kind of motion a few times in rapid succession, a cat orients itself properly and lands on its feet. There is thus a way of cheating, as it were, round the law of conservation of angular momentum in a way that is impossible in the case of linear momentum.

To sum up all these considerations about linear and angular motion: There is a state of no rotation but *no state of rest as far as linear velocity is concerned.* A system that is not rotating and not accelerating is called an *inertial system,* and there exists an infinity of inertial systems, any two inertial systems moving relative to

---

the account a little simpler, only the fundamental features of the necessary motions will be discussed.

each other with constant velocity in a straight line and without any rotation whatever. It is this large variety of inertial systems that is one of the deep consequences of Newtonian physics and, as we shall see, this is carried over without any alteration into relativity.

CHAPTER IV

LIGHT

We shall now consider another branch of physics, but starting from a different set of experiences, those connected with light.

There are two properties of light that are immediately apparent from daily experience. One is the fact that light travels in straight lines and the other is the fact that it travels at very high speed indeed. As for the first, we all know that we cannot look around corners. As for the second, light travels so fast that we are not aware in everyday life of its taking any time at all. On the basis of these two properties and related ones (reflection, refraction), a subject known as geometrical optics is built up. Highly useful, it is yet in many ways incomplete and wholly divorced from the rest of science. A physical theory that links light to other parts of physics is relatively recent and is due essentially to Maxwell.[1] The first experiment that displayed quite unambiguously such a link between light and something else (magnetism, in this case) is due to Faraday.[2]

[1] James Clerk Maxwell (1831–79), a Scot, is one of the great figures of physics. His famous equations described the electromagnetic nature of light and predicted the existence of radio waves. They are often referred to as the only laws of physics that have withstood every assault from the twentieth century's developments in scientific thinking.

[2] Michael Faraday (1791–1867) was a great English experimenter who discovered electromagnetic induction and developed

FARADAY AND THE POLARIZATION OF LIGHT

Light can be given a property known as polarization. We need not consider this phenomenon in any detail, but its chief character is that of a combing of light. When light is passed through particular media like certain crystals, it emerges as though it had been combed

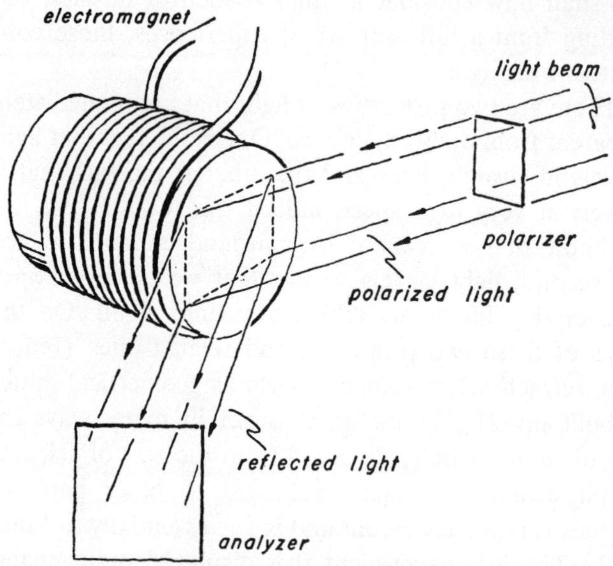

FIG. 5. *With the magnet turned off, polarized light is reflected from the metal and transmitted by the analyzer, showing that its plane of polarization is unchanged. However, with the magnet turned on, no light is passed by the analyzer; the plane of polarization has been altered.*

---

the concept of "fields" to replace the idea of action-at-a-distance. Maxwell formulated his mathematical equations to describe Faraday's discoveries and concepts. Faraday's achievement was the more remarkable in that he was self-educated.

in a direction at right angles to the direction of propagation. This direction of combing is called the plane of polarization. If the crystal is turned, then the plane of polarization of the emerging light has been turned, too. The plane of polarization is something that, in general, is rather well fixed, once the light has been polarized. For example, the light cannot then pass through a like crystal at right angles to the first one, simply because it is combed in the wrong direction to go through that crystal. Faraday discovered that although the plane of polarization of light is not changed by reflection from an ordinary mirror or metal surface, it is turned if it is reflected from a powerful magnet (Fig. 5). In this way, Faraday could show that there was a link between magnetic forces and light, and so he paved the way for the later work of Maxwell, which displayed light as a phenomenon of electromagnetism.

Maxwell and the Electromagnetic Theory of Light

What Maxwell found was that changes in the electromagnetic field, as it is called, travel with a very definite velocity that could be inferred from experiments on electromagnetic induction in the laboratory, a speed which turned out to be equal to the speed of light. The coincidence between this speed inferred from laboratory experiments and the measurements of the speed of light is a powerful argument in favor of the electromagnetic theory of light. The most startling result of Maxwell's theory is that light is only a rather special case of all these disturbances, which are all wave phenomena. That is to say, they all tend to have a period and a wave length, and what Maxwell showed was that such disturbances would travel with the speed of light whatever the wave length. This paved the way for the discovery by Hertz[3] that ordinary electric disturbances, such as discharges, produce an electric field some distance away through the propagation discovered by Maxwell. From here it was a short step to the detection and transmission of radio waves with their enormous variety of wave length, from the very long waves used in wireless telegraphy to the short waves used in TV and radar. Thus ordinary electrical apparatus can be used to transmit and receive radio waves varying in wave length from perhaps a tenth of an inch to the wave lengths of a few yards or meters used for TV, and on to wave lengths of many, many miles used in wireless telegraphy. Corresponding to the wave length there is the frequency, that is, the number of oscillations per second. Frequency is measured in cycles or thousands of cycles (kilocycles) or millions of cycles (megacycles). For waves yet shorter than the shortest radio waves, it is not electrical apparatus but atomic or molecular excitations that are used, and for the very shortest wave lengths nuclear excitation. It so happens that the retina of our eyes contains atoms of materials that respond to a particular range of wave lengths—a range of wave lengths centered around one fifty-thousandth of an inch. The longest of these wave lengths excites those particular atoms that give us the sensation of red, the intermediate ones go through yellow and green to blue and violet, which are the very shortest ones. Actually, the mechanism of color vision is very complicated and cannot just be represented in the manner of particular wave lengths producing particular colors; it is a far more

[3] Heinrich Hertz (1857–94) was a German physicist best known for his experiments on electromagnetic waves, but he also did first-class work in other areas of physics and wrote an important book, *Principles of Mechanics.*

complicated system in which the whole picture has to be taken into account.

Longer than visible light in wave length but shorter than radio waves is a type of radiation that is called infrared or heat rays, whereas rather shorter than visible light there is ultraviolet, some of which is responsible for browning our skins and giving us a tan. Still shorter wave lengths are called X-rays, and even shorter are the gamma rays that occur in nuclear processes. The enormous power of Maxwell's theory is shown by the fact that it can comprehend this tremendous range of waves, differently excited, differently received, and yet all traveling in accordance with the laws that he found. The law of rectilinear propagation applies to all these waves, but the wave character of the motion means that the waves can flow a little bit round objects that are small compared with the wave length. For radio communication round the globe one does not, however, rely on this property but on the fact that owing to peculiar radiations received from the Sun a substantial layer of the high atmosphere acts as a mirror for radio waves. This layer is the so-called ionosphere, and it is a most convenient fact that waves of more than about 15 meter wave length are reflected by the ionosphere and are thus trapped between the surface of the Earth and the higher layers of the air. Owing to this fact we can have radio communication all round the world. On the other hand, for some purposes, such as TV, higher frequencies are required, essentially because the lower ones are too sluggish to convey the tremendous amount of information that goes into making up a radio picture. Therefore TV reception is limited, more or less, to the areas from which reasonably straight paths can be run from transmitter to receiver, and therefore the coverage for television is much more difficult to achieve than for long-wave radio. For the best reproduction of sound, too, there are advantages in using these very short waves and this is the manner in which frequency modulated high-frequency radiation is sent out to give us the best possible reception. The waves employed there are long enough to flow round minor obstacles like trees and, to some extent, even buildings.

Using Radar to Measure Distance

A special application of short radio waves that is of great importance in war and peace and of considerable interest to the physicist is radar. As is well known, the notion of radar is that you transmit a short pulse of radiation, which bounces back from the target; and the wave received gives one information about the distance and direction of the target, the direction being simply that in which the wave had to be sent out to be received back. Of particular interest to us here is the manner in which the distance of the object is found. What one does is to measure the interval of time between the transmission and reception of the pulse. Since one knows that radio waves travel with the speed of light, this interval multiplied by the speed of light gives one the total length of path traveled by the packet of waves, which is to the target and back, and thus twice the distance of the target. The great interest of the principle of this method of measuring distances is that one does not use a yardstick. No standard meter or standard yard is employed. What one does is to measure an interval of time and then multiply this by a quantity, the velocity of light. We can then speculate a little about the true nature of distance. It is often very useful in physics to try to get away from the particular circumstances of our existence, in which certain materials are cheap and others expensive, in which the temperature variations are small round a mean temperature that is far removed from the very cold of space or the very hot of the stars, and so on. In the present instance, we want to make only a very simple assumption. We shall suppose, for the moment, that the makers of radar sets have managed to become so much more efficient than the makers of yardsticks that instead of rulers one normally employs a radar set, measuring the time between transmission and reception of a pulse. If, then, we had grown up measuring distances through the measurement of times, if we had grown up without using inch tapes and the like, then I should guess that the whole notion of a scale of distance would not occur to anyone.

The Units of Distance

One would always use time to express distance. This of course is done to a considerable extent in astronomy, where, to get away from the awkwardly big figures arrived at otherwise, we measure the distances of stars in light years, that is, the distance light travels in a year. But there is no reason why this manner of expressing distances should be confined to the very large astronomical distances. We could speak of light microseconds, that is, the distance light travels in a microsecond—a millionth of a second. This distance is about 300 meters or 330 yards and is quite a convenient unit. A light millimicrosecond would then be a thousandth of this unit and would roughly equal a foot in length, and so on. If we imagine, then, a civilization in which the yard or meter is unknown and every distance is expressed as light second or light millimicrosecond or whatever the case might be, then the members of that sort of civilization would look at one very blankly if one asked them what the velocity of light was. They would not regard this as a quantity to be expressed in meters a second or miles a second, as the case may be, but simply as a unit, the natural unit of velocity. Velocity one would refer to as an object moving as fast as light. All ordinary velocities would be expressed in terms of this standard. Thus the velocity of a jet plane would be around one-millionth; that is to say, a jet plane takes a million times as long to get from one place to another as light does. Similarly, a train or fast car might have a velocity of one part in ten millions (67 miles per hour approximately). In other words, by accepting the natural standard of velocity, the velocity of light, this civilization would have done away with the need to register both a standard time and a standard distance and to use the awkward translation figure of the velocity of light. There would be only a time standard in this civilization which would make life rather more convenient, and its members would regard us as people who work with distances and times in a most complicated and absurd way.

Perhaps it might be worthwhile here to describe an alternative civilization, which has the same relation to ours as we have to the one just imagined. This is a civilization in which the direction north and south is regarded as holy and is measured always in miles, whereas the direction east and west is regarded as perfectly ordinary and profane and is always measured in yards. If people had been brought up to look at things that way from an early age, it would require a daring mind to suggest that there was some connection between distances north and south on the one hand, and east and west on the other, and physicists would be employed to work these things out and would then arrive at the remarkable result that 1760 yards of the east/west measure were, in some sense, equivalent to the mile of the north/south measure and this figure of 1760 would acquire a sacrosanct meaning there, a little as the velocity of light has for us. Of course, we must imagine that in this civilization their national physical laboratory would keep two quite different standards–a standard mile for measurements of north/south directions and a standard yard for measurements of east/west directions. This would look absurdly complicated to us and unnecessary, but this, I am sure, is what we would look like to the civilization about which we talked earlier. The physicist has no hesitation in jumping to the simpler conclusion, and he immediately agrees then that there is really no point in using a standard of distance; all we need are standards of time. Again, there is the question as to what the value of the velocity of light is. The velocity of light, of course, is unity by definition. What we call measuring the velocity of light would seem to him a complicated and roundabout way of determining the length of the standard meter in Paris in terms of the well-known public standards of time.

The Velocity of Light

On this basis, then, we can look at time standards as primary and at distance standards as quite secondary and of little importance. This does seem to be a sound procedure, particularly when we think about what our inch tapes and measuring rods are actually made of. We know that they are composed of atoms whose structure is kept in shape by electric forces. We know that these atoms have certain periods of vibration and we know that, in the materials we call very rigid, it is as a consequence of the particular periods of vibration of the atoms that different atoms keep a definite distance apart in the structure of the rod. Thus we can argue that the length of a rod is really determined by the period of the oscillation of the atoms of which it is composed, this being translated, in the usual way, through the velocity of light into distance. If we argue, as we well may, that the distances between the atoms in what we call rigid materials are the distances corresponding to the oscillations of the atoms, then we could say that those distances, too, are effectively determined by radar methods. On that basis, then, distance becomes a purely secondary quantity, time is the primary thing, and the velocity of light is in natural units necessarily equal to unity. But if we are so perverse as to choose to measure distances in feet rather than in light millimicroseconds, then we have to introduce a conventional factor of conversion which effectively defines the foot, and this we call the velocity of light.

CHAPTER V

PROPAGATION OF SOUND WAVES

For our purpose the most interesting properties of light concern its propagation. In order to bring out these properties it may be worthwhile to consider the contrast with other wave phenomena. The most familiar of these is sound. Sound consists of pressure variations in the air traveling at the speed of sound, which is around 370 yards a second or about 750 miles per hour. The wave length (and with it the frequency) of sound determine the pitch of the sound we hear. A low note corresponds, in extreme cases, to 20 cycles or thereabouts, which is to say, a wave length of some 18 yards. (The product of wave length and frequency equals the velocity of sound.) Similarly the highest notes we can perceive (which vary with individual and age) are around 20 kilocycles, which is a wave length of perhaps half an inch. The usual register of voice occupies a band centered on one kilocycle, that is wave lengths of the order of a foot. It is easy to make a wide range of observations on the velocity of sound owing to the fact that it is so much lower than the speed of light. Thus, for example, if one observes a man hammering at a considerable distance one can perceive the difference in the arrival time of light and of sound, the light requiring only a negligible time–microseconds–to cover the distance, whereas the sound wave may take several seconds (Fig. 6). Thus it is quite a simple matter to establish the properties of the propagation of sound.

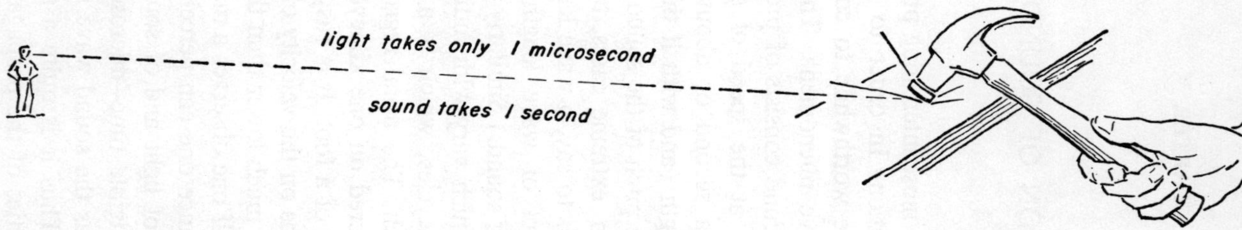

FIG. 6.

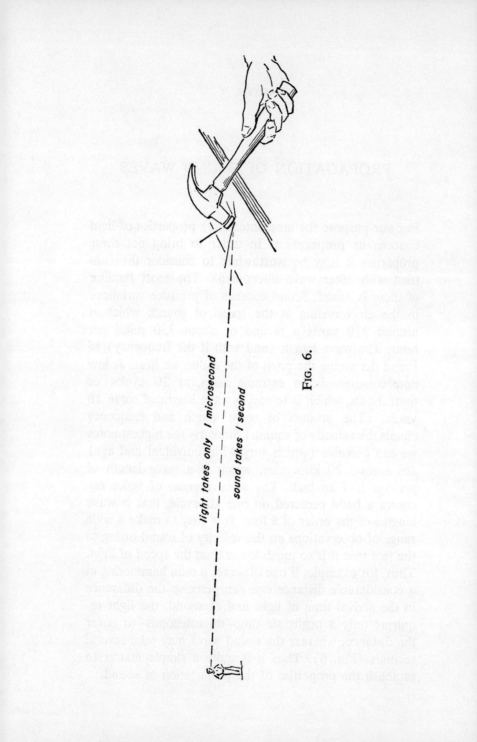

The Doppler Shift

One of the most interesting of these is the so-called Doppler[1] shift, a fairly familiar effect. If a hooting railway engine passes one at high speed, then the pitch of its whistle seems to drop suddenly when the train is at its closest. This is clearly an effect due to the motion of the engine, for if both the source of the sound and the listener are at rest in the air, then the listener receives sound of precisely the same pitch as emitted. It is not difficult to see how this occurs, but perhaps the argument will be made simpler if we consider a very fast-moving source of sound, say one traveling at half the speed of sound. For the purpose of this argument we take the velocity of sound to be exactly 370 yards per second (757 miles per hour). Suppose then that the source of sound is moving toward us, with ourselves, the receiver of the sound, at rest in the air. Consider this source of sound at two instants one second apart. At the second instant it is 185 yards closer to us than at the first instant. The sound coming from it at the second instant has 185 yards less to travel than the sound emitted at the first instant. The sound takes just half a second to cover these 185 yards. Therefore the second sound has a travel time half a second less than the one emitted at the first instant and so, since they were emitted one second apart, the sound from the second moment arrives only half a second later than the sound from the first moment. Instead of considering moments one second apart we can consider them a very much shorter period apart. We can consider, say, the high-pressure points of a sound wave of a thousand

[1] Christian Johann Doppler (1803–53) was an Austrian.

cycles emitted by the source. Then the peaks of the wave occur a thousand times a second, and are thus a millisecond apart. In that millisecond the source will have traveled .185 of a yard and the sound after that millisecond will, therefore, take half a millisecond less to come to us. Therefore, the pressure peaks when they arrive at us will appear to be only half a millisecond apart corresponding to a frequency of 2000 cycles per second. This means that we will perceive the pitch of the sound as one octave higher than when it was emitted by the source. It is not difficult to see that when the source is receding from us then each successive sound wave has a greater distance to cover, and so the interval between the arrival of successive peaks is higher than the interval between their emission and, accordingly, the pitch of the sound will be lowered. This accounts for the well-known effect of the railway engine and is referred to as the Doppler shift. What is not quite so well known, and interesting in view of what we shall do later, is that the Doppler shift is not the same when the source is moving relative to the air with the receiver at rest in it (as in the case just discussed) as it is when the receiver is moving through the air with the source at rest in it. Suppose we are on a fast vehicle traveling at half the speed of sound toward a source of sound. Suppose, again, that we consider the noise emitted by the source at two instants a second apart. At a particular moment we will receive the noise emitted at the first instant. Then consider the situation two-thirds of a second later. Since we are traveling with half the velocity of sound (185 yards per second) we will then be 123 yards or so nearer the source. Sound emitted by the source will, therefore, take one-third of a second less to reach us than when we received the first noise. The noise emitted one second after the first emission considered will reach us therefore at this moment, two-thirds of a second after the first noise was received. For though it started on its journey a whole second later, the diminution of distance reduces its journey time by just a third of a second. Thus, if we are moving toward the source, then noises emitted at intervals of one second will be received by us at intervals of two-thirds of a second. Hence, the frequency of the noise is increased by 50 percent now, not by 100 percent as in the case when the source was moving, and so the pitch instead of being raised one octave is only raised by a fifth. It is, therefore, not enough to consider just the velocity of the source relative to the receiver; one must consider both the velocity of the source relative to the air and the velocity of the receiver relative to the air. For small velocities it does not make much difference which is moving but, when one considers, as just now, velocities as high as half the velocity of sound, then it matters quite appreciably.

The Sonic Boom

These effects, striking enough at moderate speeds, become quite extraordinary when speeds higher than the velocity of sound are considered. With the advent of supersonic aircraft this has become quite familiar in recent years, and one of the best-known and most deplored effects of supersonic flying is the supersonic bang (in America, the sonic boom) or, as it is often called, breaking through the sound barrier. As we shall see, this is a very bad name indeed, but the phenomenon itself is interesting and instructive.

Consider first a plane flying, as in Fig. 7, on a straight level course at a height of two miles (10,560 feet) at a speed in excess of the velocity of sound, such that

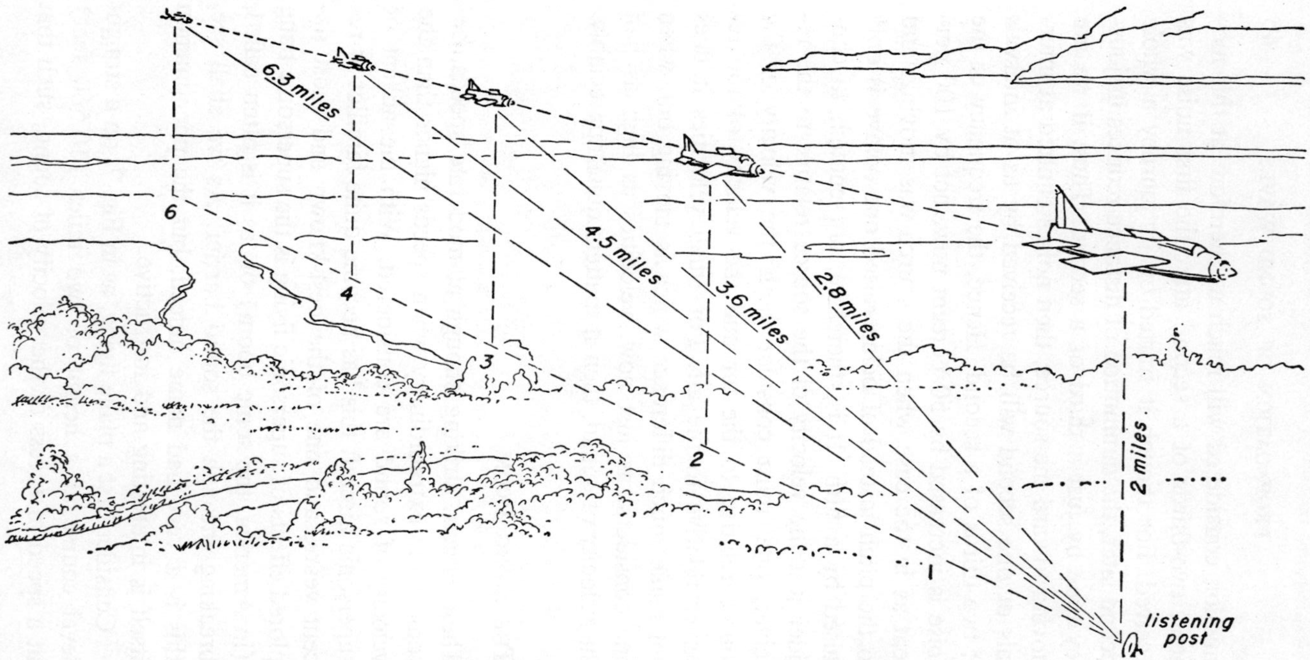

FIG. 7. *The Boom of a Supersonic Aircraft*

it covers a mile in every four seconds, whereas we may take for the purposes of this argument that sound covers a mile in every five seconds (720 m.p.h.). When the plane is far away, then the part of its speed that is directly toward us is greater than the velocity of sound. In other words, the plane diminishes its distance from us more rapidly than a sound wave does, and therefore we receive the sound from a later moment in the flight of the plane earlier than we receive the sound from a preceding moment in the flight of the plane. However, when the plane gets more nearly overhead, its distance from us does not diminish nearly so rapidly. In fact, the distance of the plane from us reaches a minimum when the plane is directly overhead and after that increases again. As soon as the main reason for the distance of the plane, as it were, is its height, which is not changing, the speed of the plane does not imply such a rapid diminution of its distance from us. It then stays more nearly at the same distance. If its distance diminishes less rapidly than the velocity of sound, then sound from later moments in the course of the plane will reach us later, just as ordinarily. Thus the distant parts of the flight we hear, as it were, in reverse and the later parts of the flight we hear in their correct sequence. There is, therefore, a first moment at which we hear the plane at all and then noise from quite a stretch of the flight of the plane arrives simultaneously, because the distance of the plane from us is then diminishing at just the rate at which sound covers distance. Therefore the noises emitted at successive moments all arrive at the same time. It is this simultaneous arrival of noise from quite a stretch of the flight of the plane that through the summation of the noise produces the bang.

Suppose, to make it more definite (Table I), that,

when we start our stopwatches at zero hour, the plane is at such distance from us that the point directly under the plane is exactly 6 miles from us. Owing to the height of the plane its distance from us is a little larger, say 6.3 miles, and sound will take approximately 31.5 seconds to reach us from there. Eight seconds later the plane has covered 2 miles, the point directly below the plane is thus only 4 miles from us, and the distance from the plane to us is about 4½ miles, which sound covers in 22½ seconds. Since the sound started 8 seconds after the sound from the previous moment considered, it arrives 30½ seconds after zero and thus 1 second *before* the noise emitted 8 seconds earlier. Four seconds later the distance of the point just below the plane from us is only 3 miles, the distance of the plane from us is 3.6 miles, which sound covers in 18 seconds, and the sound arrives half a second earlier than the sound emitted 4 seconds before. Four seconds afterward the plane is now at an elevation of 45 degrees so that a point directly under the plane is only 2 miles from us and the distance from the plane to us is about 2.8 miles, covered by sound in approximately 14 seconds. Therefore the noise from this moment arrives at precisely the same instant as the noise emitted 4 seconds earlier. The earliest sound, in fact, arrives a shade before this from a point in the flight of the plane between the last two positions considered. Four seconds later the point directly below the plane is only 1 mile from us, the distance of the plane is now 2¼ miles, covered in about 11 seconds by the sound, which therefore arrives a second later than the sound emitted 4 seconds earlier. Four seconds later still, the plane is directly overhead, the sound requires 10 seconds to cover the distance of 2 miles representing its height, and thus this sound arrives 3 seconds later than that emitted 4 sec-

TABLE I

Time of Emission of Sound (seconds)	*Ground Coordinate of Aircraft (miles)*	*Slant Distance between Aircraft and Listener (miles)*	*Travel Time of Sound (seconds)*	*Time of Arrival of Sound (seconds)*
0	6	6.3	31.5	0+31.5
0+ 4	5	5.4	27.0	0+31
0+ 8	4	4.5	22.5	0+30.5
0+12	3	3.6	18.0	0+30
0+16	2	2.8	14.0	0+30
0+20	1	2.25	11.25	0+31.25
0+24	0	2.0	10.0	0+34

onds earlier. Thus, until nearly 30 seconds after the start of our investigation, we hear nothing at all. Then suddenly we hear, as a bang, the noise from quite a period in the life of the plane, and after this we hear simultaneously, going forward in time, the noise emitted by the plane after this instant and, going backward in time, the noise emitted by the plane earlier on when it was farther from us. Thus the noise emitted by the plane when it was 4 seconds after the beginning of our investigations and the point below the plane was 5 miles from us arrives at precisely the same moment as the noise emitted by the plane when the point just below the plane was 1 mile from us.

Another kind of phenomenon also occurs frequently. Suppose the plane was originally flying on the same course as before and at the same level but at slightly less than the speed of sound. Thus although it was approaching us fairly fast, the noise from successive moments of the plane's flight was received by us in that order. Then suppose that the plane speeded up to above the velocity of sound until the start of the previous investigation is reached, when, as will be remembered, the noise of the plane was received backward in time.

Between the going forward of the subsonic part of the flight and the going backward of the supersonic part of the flight there must have been a moment when we received sound from a whole stretch of flight simultaneously—another bang. This will have occurred at a time when the plane was diminishing its distance from us at just the speed of sound. It would have had to go at somewhat more than the speed of sound at that time, since in level flight it could not have come directly toward us. Thus in such a flight there are two supersonic bangs, and the first one discussed here is the first one received. Before this bang we hear nothing at all; after this bang we hear simultaneously noise from three periods in the flight of the plane: the subsonic period before it increased its speed and the noises emitted then we hear in the right order; the part of the supersonic portion of the flight far from us we hear running backward in time, and the later part of the flight when it was more nearly overhead and the noise again arrives in the right order in time but simultaneously with the other two lots. A little later we will hear the second bang—the one discussed second in our considerations; after this we only hear the noise emitted by the plane when it is overhead or flying away.

The difference between the motion of the receiver and the transmitter is shown by the fact that supersonic bangs can only occur if the *transmitter* is moving at a speed exceeding the velocity of sound. If a plane is flying toward a source of noise on the ground at a speed exceeding the speed of sound, the people on the plane will not hear any form of supersonic bang. For a bang is heard only if noise from a whole period arrives simultaneously at the receiver. With the transmitter at rest in the air sound waves would have to overtake each other to produce this effect, and this cannot happen since the speed of sound is the same for all sound waves and so no bang is perceived.

Another familiar class of waves is that on the surface of water. Like sound waves they travel in a medium, the water, but their properties are rather more awkward, since the speed of a water wave depends on its wave length. Thus waves 500 yards from crest to crest rush across the ocean at almost 60 m.p.h., while waves measuring only one foot from crest to crest move at a mere 1.5 m.p.h. Nevertheless many of the features of sound waves discussed in this chapter can be illustrated with water waves.

CHAPTER VI

THE UNIQUENESS OF LIGHT

When Maxwell showed, about one hundred years ago, that light consisted of waves, people naturally turned to other types of waves for analogies so as to gain insight into the phenomenon of light propagation. There are many such other kinds of waves—sound waves and water waves, waves on stretched strings, earthquake waves, and so on.

All these waves travel of necessity through a medium, which may have a velocity of its own that can be ascertained quite independently of wave propagation and, moreover, the velocity at different points of the medium need not be the same, leading to complicated processes of scattering and refraction. It is perhaps not surprising that at first people did not fully appreciate how radically different the propagation of light was from all these other phenomena. They invented a hypothetical medium for the propagation of light which they called the ether. As so often happens in science, this was an idea, conceived on a nonexistent analogy, that misfired completely and confused instead of helping.

A Hypothetical Ether

The ether served one purpose and one purpose only —to account for the propagation of light, to be for light what air is for sound. But air can be weighed, it can be pushed around, it can be pumped out or it can be put under pressure. None of these things can be done with this hypothetical ether. The ether must be everywhere because it cannot be removed. Ether cannot be pushed around; otherwise it would exert a retarding influence on the planets and introduce friction into the Solar System where the motions of the planets show that there is no such friction. The ether is not moved even by big bodies, for when the Moon passes in front of a star, the light from that star is received without any observable change whatever until the Moon obscures it. This means that the Moon, even directly above its surface, exerts no effect on the ether. Thus, the ether has no properties bar one: it helps to make an analogy between the propagation of light and the propagation of sound. But this is readily seen to be a false analogy by considering Newtonian dynamics.

We have pointed out that uniform motion does not affect dynamical processes. You will recollect that attention was drawn to the complete and exact likeness between pouring out a cup of tea in an airliner and pouring out a cup of tea at home at rest. This likeness was referred to as the Newtonian principle of relativity. All inertial observers are dynamically equivalent in the sense that if each of them carries out an experiment in his own surroundings he will get precisely the same answer as any other inertial observer. To put it a little differently, we found that velocity was irrelevant, and only acceleration mattered. For example, in a comfortable airliner we can talk to our neighbors just as easily as we can talk to them when we are both on the ground. This is a simple example of Newtonian relativity. Sound is essentially a dynamical phenomenon, concerned with the motion of the air. The speaking person (transmitter), the medium of propagation (the air), and the listening person (receiver), which are all that matter for this consideration, all move together with the speed of the plane and, therefore, have the same experiences as the people on the ground.

The Absurdity of the Ether Concept

What happens to light on the plane? The only simple way–indeed, to us nowadays the obvious way–is to say that just as one can talk as easily on a plane as on the ground, so one can read as easily, although this involves the propagation of light. But according to the ether concept, absurd as this seems to us now, the plane would have been in a rather different position from the ground. As the evidence of the Moon showed earlier, the motion of objects cannot drag the ether along with them. Accordingly, if we suppose the ether to be at rest relative to the people on the ground, there would be an ether wind blowing through the plane, and this would affect the propagation of light. It is true that the velocity of the plane is very small compared with the velocity of light (barely one-millionth of it), but the same arguments would apply to far higher velocities. Furthermore, instruments for measuring properties of light are exceedingly sensitive. The concept of the ether, therefore, involves the absurd consequence that by *optical* means one should be able to distinguish between being in a state of uniform motion and being at rest, although it is impossible to do so by dynamical means. This misconception is clearly contrary to all the ideas that were stressed earlier concerning the unity of physics; the impossibility of separating optics and dynamics and other branches of physics. If there is no way of distinguishing between different inertial observers by dynamical means, there cannot be any method of distinguishing between them by any other means.

This clear and obvious extension of the Newtonian Principle of Relativity was not accepted seventy years ago. So powerful had the ether concept become that people went about "measuring" this ether wind. Now it is clear to us that the whole ether concept in suggesting an ether wind was harmful and misleading, but this was very hard to see at the time. One of the most famous experiments in physics, the Michelson-Morley experiment, was undertaken to find this ether wind, and only its failure led to the recognition of the fact of the unity of physics, the notion that if velocity did not matter for dynamics it could not matter for optics either. Led on by the absurd ether concept, people realized that one could not expect the ether to be standing still relative to the Earth, which is rushing round the Sun at a speed of nearly twenty miles per second, roughly one part in ten thousand of the speed of light. But the Earth cannot push the ether round with it. There is the evidence that the Moon cannot do it; moreover, a motion of the ether pushed or dragged along with the Earth relative to the ether farther away would show itself in complicated refraction effects giving the stars an apparent motion that is not, in fact, observed. Thus an ether wind should be expected blowing at about one ten-thousandth of the velocity of light. How would one measure this ether wind? It would show itself as a difference of the velocity of light in different directions.

Measuring Velocity

The usual method of measuring the velocity of an object is to measure a distance, and then to station observers with clocks at each end to time the passage of the object. The distance divided by the difference in the clock readings is then the velocity. For a given distance, the faster the object the more critical will be the clock readings, and even a small error in the synchronization of the two clocks will completely invalidate the determination of the velocity. For anything as fast as light this is clearly a very serious difficulty over any practicable distance between the observers, particularly if one is trying to measure the speed of light with sufficient accuracy to notice the effect of the ether wind on it. There is the additional difficulty that the observers normally would synchronize their watches by signaling to each other with light, therefore making the synchronization dependent on the very thing they are trying to measure —viz., the speed of light. These difficulties are avoided by working in one place only, and measuring the length of the round trip of light from the observer to a mirror and back. One therefore would compare the timing of an upwind/downwind round trip with that of a cross-wind round trip. An example may help to make the argument clear.

Suppose we are on a river two miles wide that flows at a speed of three miles per hour (Fig. 8). We have a boat capable of five miles per hour in still water. We want to use it on one occasion to visit another place on our shore two miles upstream from us and return, and on another occasion to cross the river to the point opposite us and come back. How long will these expeditions take? First, traveling upstream our boat will travel at effectively two miles an hour using its power velocity of five miles an hour against a stream of three miles per hour, so that we shall take one hour to reach this place upstream. On the way back the velocity of the river will be with us and, adding our five miles an hour to the river's three miles an hour, we travel back at

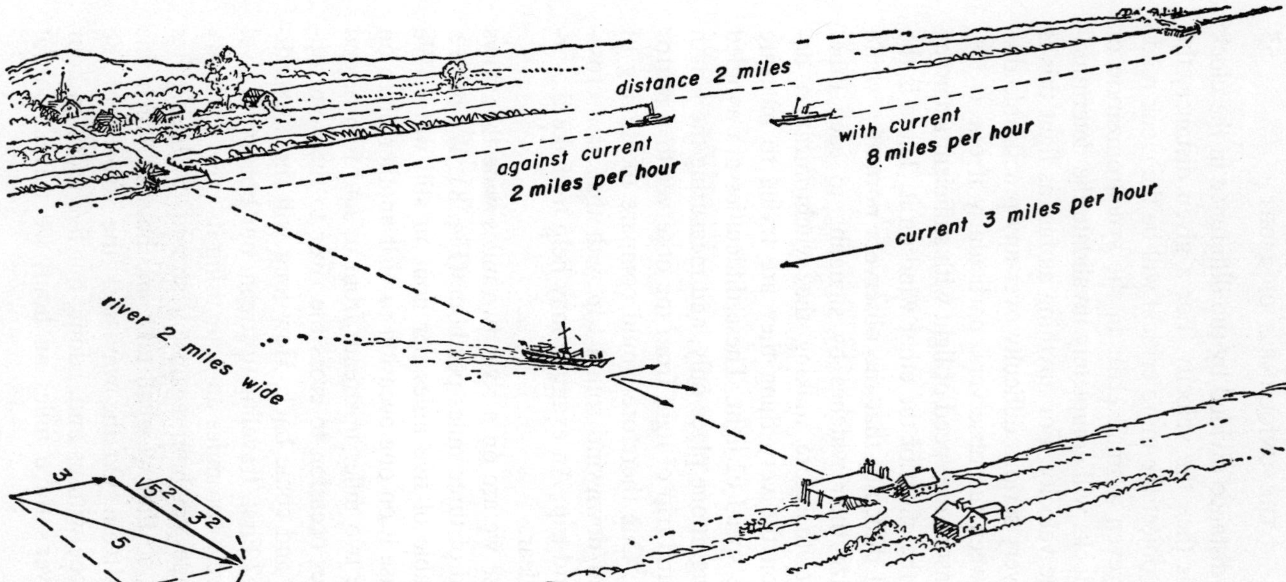

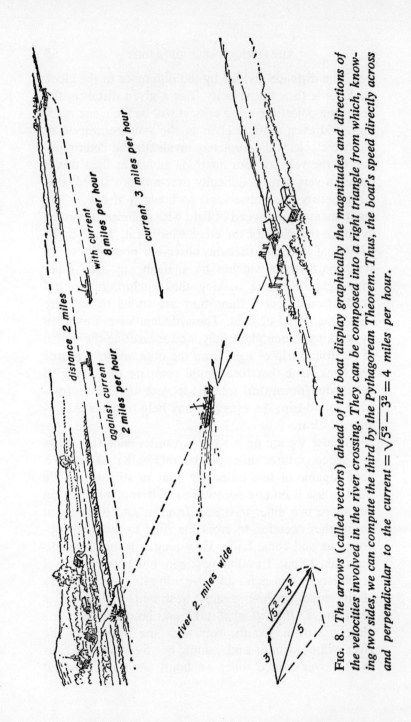

FIG. 8. *The arrows (called vectors) ahead of the boat display graphically the magnitudes and directions of the velocities involved in the river crossing. They can be composed into a right triangle from which, knowing two sides, we can compute the third by the Pythagorean Theorem. Thus, the boat's speed directly across and perpendicular to the current* $= \sqrt{5^2 - 3^2} = 4$ *miles per hour.*

eight miles an hour, covering the two miles in a quarter of an hour. Therefore the trip there and back will take one hour and fifteen minutes. Crossing the river, we must not point the nose of our boat straight across because then we would be drifting downstream to a place much lower down the river. We must turn the nose of our boat partly upstream, so that while we are going five miles an hour in the direction the boat is pointing, three miles per hour of this is an upstream direction to counteract the velocity of the river. Looking at the drawing (Fig. 8) it will be seen that this leaves us an effective velocity of four miles an hour for crossing the river so that it takes us half an hour to get across and half an hour to get back, just one hour altogether. Thus traveling to a place two miles away and coming back to the starting point takes different times whether the journey is upstream and downstream, or across the stream. This was the method used in the celebrated experiment of Michelson and Morley.[1]

The Michelson-Morley Experiment

Light was sent off in two directions and reflected from places the same distance from the starting point and then the travel times, or, to be precise, the number of wave lengths, were compared. Taking the speed of the Earth as approximately 20 miles a second, roughly one part in ten thousand of the velocity of light, the dif-

[1] Albert Abraham Michelson (1852–1931) was the first American scientist to receive (1907) a Nobel prize. Besides contributing to the development of relativity, he measured the speed of light and invented an interferometer for precise measurement by means of wave lengths of light waves.

Edward W. Morley (1838–1923), Michelson's collaborator in the famous experiment on ether drift, was a chemist by training. See *Michelson and the Speed of Light* (Science Study Series).

ference in the travel times should be only one part in two hundred millions, but so accurate are the methods of spectroscopy that this difference is detectable by so-called interference methods in which waves returning from the two directions are combined. First the pattern made by the combination of waves is noted for one position of the apparatus which is then turned through an angle. If this change of direction changes the travel times of light (through the cross-stream arm becoming the upstream-downstream arm and vice versa), then this would show itself in a shift of the interference pattern of the two waves. No such shift was observed. So persuasive was the pernicious ether concept that at the time this was regarded as unintelligible. Nowadays this negative result is clear because we suppose that the travel times of light are independent of direction, independent of whether the beam travels in the direction of the Earth's motion or opposed to it or whether it travels at right angles to it, because velocity matters as little for the propagation of light as it does for dynamics. The negative outcome of the Michelson-Morley experiment turned out to be fundamental, leading to the insight that velocity does not matter for optical phenomena any more than for dynamical ones, a statement known as the Principle of Relativity, which was first clearly enunciated by Einstein in 1905 following earlier statements by Lorentz[2] and Poincaré.[3]

[2] Hendrik Antoon Lorentz (1853–1928) was a brilliant Dutch physicist who developed a kind of theory of relativity before Einstein's in the course of working on Maxwell's equations to meet new experimental data. Einstein's basic mathematics for his Special Theory of Relativity was the same as the Lorentz Transformation, but he had not heard of Lorentz's prior work.

[3] Jules Henri Poincaré (1854–1912) was the leading French mathematician (and one of the world's greatest) at the turn of the century. Many scholars believe that he would have produced Einstein's Theory of Relativity if Einstein had not got there first.

A final remark about how nonsensical the basic experiment of seventy years ago appears to our present generation. Clearly, if one or both of the arms of the Michelson-Morley apparatus had changed their lengths

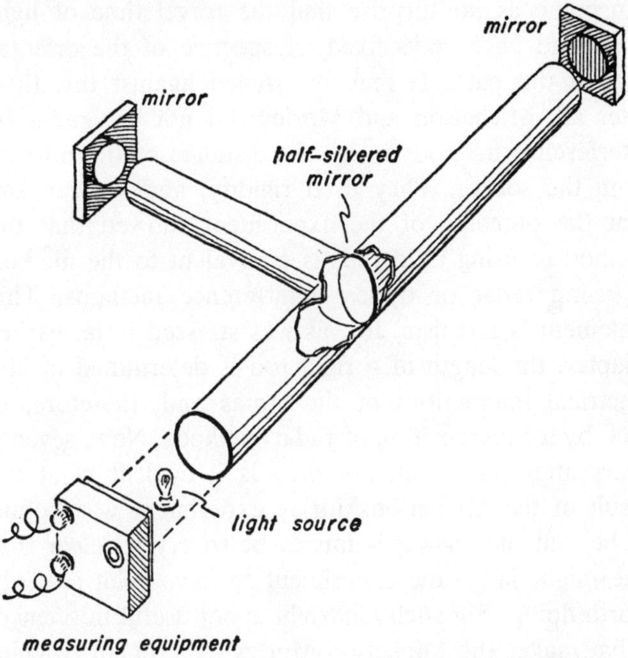

FIG. 9. *The speed of light in two different directions can be compared in an interferometer.*

in turning, then this change would have affected the result considerably. Therefore the authors of the experiment took the greatest trouble to make the arrangements as rigid as could be done. But to us now this looks very peculiar. With present-day technology we would fix the length of these arms by radar methods or by equivalent optical interference methods. But this would mean that we fix the length by measuring the time light takes to go there and come back. If we establish the distance of a mirror by the requirement that the travel time of light there and back should stay fixed irrespective of the orientation of the path, then it can come as no surprise that the travel time of light there and back stays fixed irrespective of the orientation of the path. It may be argued against this that, after all, Michelson and Morley did not use radar or interference methods to fix the distance of the mirrors from the source. They used rigidity, and we can say that the outcome of the experiment showed that the method of using rigid rods is equivalent to the method of using radar or optical interference methods. This statement is a truism, for, as was stressed in an earlier chapter, the length of a rigid rod is determined by the electrical interactions of the atoms and, therefore, in fact, by a superposition of radar methods. Now, seventy years after the event, not only is it evident what the result of the Michelson-Morley experiment was bound to be, but the answer seems to be so crystal clear that one might judge the experiment to have been scarcely worth doing. But such hindsight is not useful in science. What makes the Michelson-Morley experiment so celebrated is that it first proved something that has so deeply entered our thinking that it has become obvious. There can be no greater merit in a scientific discovery than that before long it should appear odd that it ever was considered a discovery. Only what has become obvious was really important, because only those things that have so deeply affected our thinking and so thoroughly changed our outlook that we cannot think without them have really entered the spirit of the human race.

CHAPTER VII

ON COMMON SENSE

The arguments of this book, and particularly of the last chapter, point clearly to the fact that all inertial observers should be regarded as equivalent, not only in respect of dynamics, as Newton already found, but also in respect of light. In particular, this implies that the velocity of light is the same in all directions, irrespective of the state of motion of the observer, provided only that he is inertial.

This result, as has been stressed before, is indeed obvious once it is agreed that distance measurements should be made by radar methods. For then one measures a time, half the time between the emission and the return of the radar pulse, in order to determine the distance of the target. This time, which could perfectly well itself be used to describe this distance, is, as a matter of habit, multiplied by a purely conventional (and therefore arbitrary) number, called the velocity of light, merely to express the distance in miles (or centimeters), etc. Also, the principle of the unity of physics requires that systems that cannot be distinguished by internal *dynamical* experiments should be indistinguishable by *any* internal experiments. And so we are driven, virtually without means of escape, to Einstein's Principle of Relativity: All inertial observers are physically equivalent and no internal experiment can be devised that discriminates in any way between different inertial observers. Put more simply, it says that there is no way of appreciating the velocity of one's vehicle unless one looks out of the window, provided the vehicle is unaccelerated.

Contemplating this great principle of physics after fifty-nine years, one cannot help wondering how people could possibly ever have thought differently. To say this is no disparagement of the tremendous achievement of Einstein: on the contrary, it is the mark of a really major step in thinking that, when we have become used to it, we can no longer imagine how things were before that step. In spite of its, to us, almost obvious character, the Principle of Relativity at first had a rather rough passage. Such situations occur frequently in physics if a step is taken, logically compelling, forced upon us by experiment, and yet upsetting one of our cherished notions. This is what the Principle of Relativity did to the concept of time. As will be shown in the next chapter, it changed the concept of time in a way that seemed to be contrary to common sense.

The Experience of Everyday Life

It may be worth digressing for a moment to discuss what is meant by common sense, and whether a conflict of physics and common sense is something to be expected. Common sense refers to the tremendous amount of experience we gain in early life that tells us such an enormous amount about the world we live in and the objects that surround us. Though possibly some elements of it are instinctive, the bulk of common sense is distilled experience. The experiences we gain in those early years of life are naturally gained in our ordinary surroundings, using the tools and materials that are at hand. We do not gain any common sense appreciation of the behavior of gas at a million degrees because we do not meet that kind of thing in ordinary life. Nor do we gain any appreciation of the view of the world one would have if one were speeding through the countryside at 100,000 miles a second, simply because this does not happen to us. On the other hand, common sense is a splendid guide in that particular field on which it is based—that is, the experience of everyday life. It is the task of the physicist to go beyond this, to devise and use instruments that enrich experience, that allow one to gain knowledge of objects and circumstances one would not normally come across. One would then expect that when the physicist, in his exploration of a larger range of experience than what has gone into the making of common sense, examines his results, he may find either that the impressions of common sense apply or that they do not apply. If they do apply then it merely means that, in his extension of range of circumstances, he has not gone beyond the validity of what was acquired in a narrower context. If he does run out of this set of circumstances, then, naturally, one would not expect common sense to apply. In fact, common sense would then be an unwelcome intruder making it more difficult for us to adapt ourselves to these new circumstances than it might otherwise be.

Adaptation to a new set of circumstances is always possible. Human beings are intelligent, which is equivalent to the statement that they are flexible. Once one has met enough of a new set of circumstances, one can order them and acquire a new understanding of them, in just the same way that one can learn a new language. It so happened historically that the physicists' instruments became sufficiently powerful to outrun the range of validity of common sense at the end of the nineteenth century and in the early years of the twentieth century. Then, for the first time, results were derived clearly contrary to experience in our ordinary and very different circumstances, and a great deal of heart-searching and trouble arose. Nowadays we know that this is totally misplaced; the farther we range with our instruments, the stranger the worlds we investigate, the more different they will clearly be from what we are accustomed to. The modern physicist, to follow Lewis Carroll, is used to believing at least two incredible things every day before breakfast. To put it again a little differently, the surprising thing, surely, is that molecules in a gas behave so much like billiard balls, not that electrons behave so little like billiard balls. To come back, then, to Einstein's Principle of Relativity, its fundamental importance is that it extends the Newtonian notion of relativity to all physics. All inertial systems are equivalent in every sense of the word. It is not only a question of pouring out tea in a jetliner being the same as pouring out tea at home, but it is also true that looking into a mirror in a jetliner one sees the same as one does when one looks into a mirror at home.

Time: A Private Matter

Optics, dynamics, in fact the whole of physics is unchanged by uniform velocity. It turns out, as will be proved in the next chapter, that the common sense concept of time does not fit in with the Principle of Relativity, combined with the fact that the velocity of light is the same in all inertial systems. We have to ease ourselves out of the time concept of earlier days. It turns out that this is only seriously upset when one considers very high velocities, velocities not much smaller than the velocity of light. But, in principle, the difficulty arises at all speeds. We have acquired the knowledge that if you have a watch and I have a watch and they are made by reputable watchmakers and we synchronize them at one time, then whatever you do and whatever I do, the two watches will always show the same time. However, this is a piece of experimental knowledge acquired in a very narrow context, viewed from the point of view of physics as a whole. For neither you nor I will travel with really high speeds at any stage. We must therefore contemplate the possibility, not to put it any higher, that this permanent keeping together of your watch and my watch applies only as long as we both travel very slowly. To suggest that because our clocks keep the same time when we both travel slowly, therefore they must keep the same time when we move at velocities close to that of light, is obvious nonsense. We thus have to get used to the idea that time is a private matter. That is to say, that MY time is what MY watch tells me.

Time is that which is measured by a clock. This is a sound way of looking at things. A quantity like time, or any other physical measurement, does not exist in a completely abstract way. We find no sense in talking about something unless we specify how we measure it. It is the definition by the method of measuring a quantity that is the one sure way of avoiding talking nonsense about this kind of thing. If, therefore, we enter the discussion of the nature of time with an open mind on the basis that in everyday life we do not actually test what happens to time when we move very fast, then we must be prepared for universal public time to break up into a multitude of private times.

The "Route-Dependence" of Time

The possibility of time being private rather than public raises the question of the "route-dependence" of time. In everyday life there exist two quite different

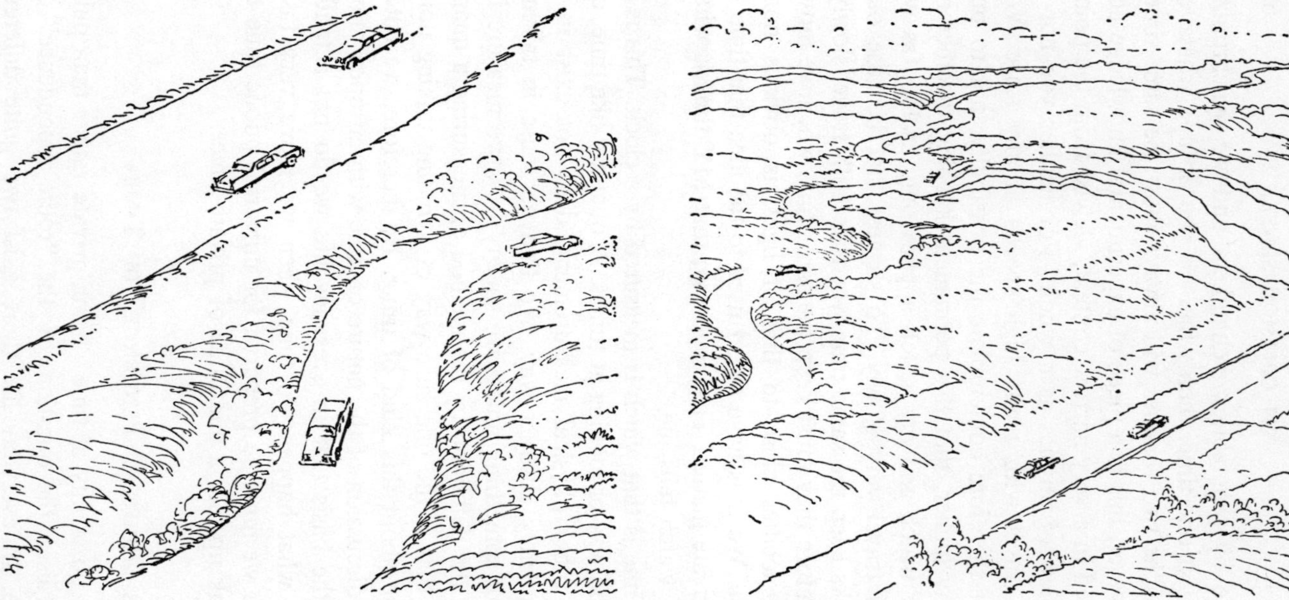

FIG. 10. *Mileage is a route-dependent quantity.*

kinds of quantities which we may describe as route-dependent and route-independent. The distinction between them is readily explained when we consider a journey across hilly country (Fig. 10). If we start in one town and drive across the hills, up and down, over various ridges, to another town, and keep account always of the gain of height and the loss of height in every part of our way, then the net amount of height we have gained is simply the difference in the levels of the town where we finished the journey and the town where we started it. Had we followed a very different route, starting and finishing at the same points, the net gain of height would have been the same, as it must be, because it is simply the difference in the heights of the starting and finishing points of the journey. This is a typical example of a route-independent quantity. The net gain of height does not depend on the route chosen. On the other hand, if we use the milometer of the car to measure the distance between the two towns, then our route will matter very much. According to the way we travel, the mileage covered between the towns may have one value or another value. The mileage is, therefore, a route-dependent quantity. This may sound a mere statement of the obvious, but the important fact is that all the quantities of physics can be classified as either route-dependent or route-independent.

Consider, as a route-independent quantity, the motion of a diesel train running a shuttle service on a single-track line between two towns (Fig. 11). If there is no wheel slip, the wheels of the train will be in precisely the same position whenever the train is standing at the same point, irrespective of how many forward and backward journeys the train has undertaken. Thus, if a chalk mark were put near the top of one of the wheels when the train was in its starting position, then, when-

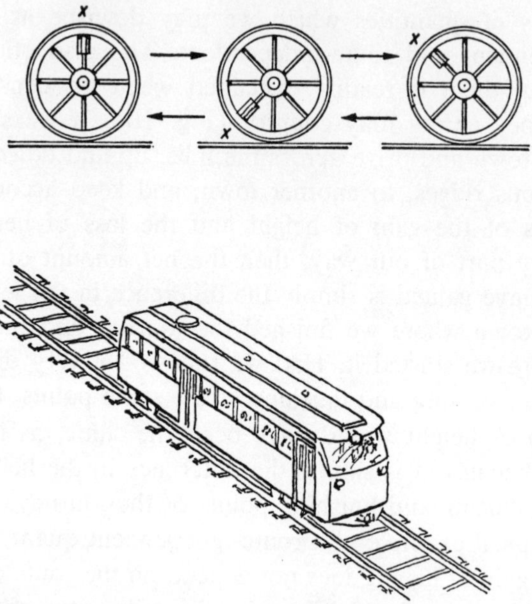

FIG. 11. *The wheel is in the same position whenever the car passes the same point, so that the rotation of the wheel is route-independent.*

ever the train returned to that position, the mark on the wheel would be near the top. Thus the position of these wheels is a route-independent quantity. We can contrast this with a route-dependent quantity such as the level of fuel in the oil tank of this train, which clearly depends crucially on how many of these return journeys it has made since the tank was filled.

How absurd it would be in ordinary life if we mixed up route-dependent and route-independent quantities is shown by the suggestion that, while it is quite reasonable for the railway to charge a substantial single fare, the return fare should always be nil, since one ends up in the place where one started. Perhaps the rule should then be that you pay a fare when you go somewhere, and when you return you get your money back. Such a route-independent charging system would bear no relation to railway costs, which are strongly route-dependent. To the physicist and mathematician there is all the world of difference between a route-dependent and a route-independent quantity.

The crucial discovery of relativity was the route-dependence of time which, having previously been considered a public universal quantity, was naturally thought to be route-independent. Of course, it is only the fact that time has become private, rather than public, that allows it to be route-dependent at all, but the important point to emerge in the next chapter is that we cannot escape within the Principle of Relativity from the notion of route-dependence of time.

CHAPTER VIII

THE NATURE OF TIME

We are ready now to demonstrate one of the most interesting consequences of relativity—namely, that time is a route-dependent quantity. In order to show this we have to consider objects traveling at very high speeds, speeds comparable with the speed of light. We do not knowingly encounter such objects in everyday life, and we never have two human beings traveling at such relative speeds. Hence we are considering strange and unfamiliar situations. But modern technology leaves us no choice. In the big particle accelerators ("atom-smashing machines") minute particles are made sometimes to travel at speeds of over 99 percent of the speed of light, and the physicist must examine situations in which such velocities do occur. He also can observe atomic particles traveling at such speeds in cosmic rays. For technological reasons, only atomic particles can be accelerated to such speeds, and the strangeness of the intrinsic properties of such particles gets mixed up with the strangeness arising from their high velocities. Our imagination, however, is untrammeled by these technological limitations; at least for the purposes of discussion we can isolate the peculiarities arising from high speeds and allow ourselves to think about human beings instead of atomic particles traveling at such velocities.

The Peculiarities of High Speeds

Since we shall need inertial observers in our discussion, we must require them to travel at constant speed. If we want to avoid thinking about terribly short intervals of time, our observers have to have plenty of room, and so we may imagine them to be space travelers, proceeding at much higher speeds than present rocket technology can give them. Nevertheless, such an image helps one to visualize the arguments to be given. These will aim at clarifying the Principle of Relativity and finally deducing observable consequences from it, so that its experimental tests can be appreciated.

It will be remembered that Einstein's Principle of Relativity asserts the equivalence of all inertial (i.e., uniformly moving) systems. Newtonian dynamics already stresses the importance of acceleration and the irrelevancy of velocity (pouring tea in a smoothly flying jetliner is no more difficult than in one's home), and so defines inertial systems. It asserts the equivalence of these inertial systems as far as dynamics goes, but earlier chapters, in stressing the unity of phyics, showed how awkward it would be if different parts of physics were governed by different rules of transformation from one inertial observer to another. Thus we were led to Einstein's principle of the *complete* equivalence of all inertial observers, and we now explore its consequence using simple ideal experiments.

It will be useful to depict our moving observers, to use the jargon of the theory, on space-time diagrams (Fig. 12). In such a diagram the vertical axis represents time, as measured by some inertial observer, and the horizontal axis represents distance. Fortunately, almost all our arguments require only one space dimen-

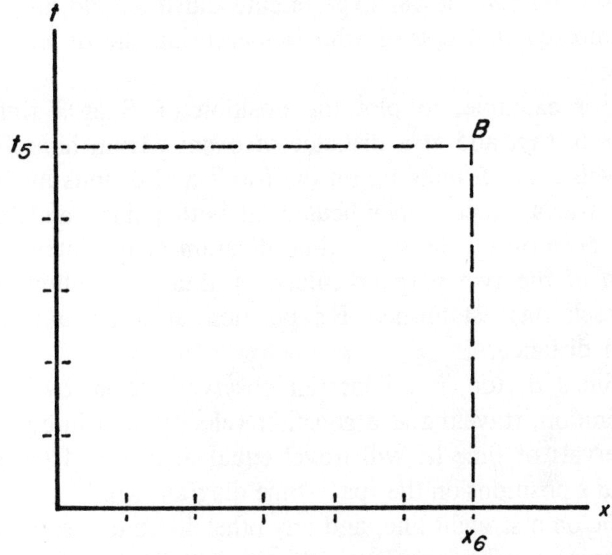

FIG. 12. *Space-time Diagram*

sion (i.e., all bodies considered move along one straight line), and so this representation is sufficient.

THE RELATIONSHIPS OF INERTIAL AND MOVING OBSERVERS

Since the observer A is the one who draws the diagram, he understandably will make himself the origin of the coordinate system—that is to say, he will begin making his observations at his own time zero, and will measure the positions of other observers (in both time and distance) from his own position at his own time zero. Because he considers himself to be at rest and thus is not going anywhere along the distance axis, his x-coordinate remains zero. Distances along the t-axis, representing change in time, show the progress of A

in the diagram of our experiment, and it should be remembered that it is *A* who is measuring the distances along *t*.

For example, to plot the position of *B* at a time 0 + 5, say, and at a distance of 6 units from himself, *A* will count 5 units up on the *t*-axis and 6 units along the *x*-axis, erect perpendiculars at both points, and fix *B*'s position on the space-time diagram at the intersection of the two perpendiculars. *A* thus in relation to himself has established *B*'s position at a given time and distance.

Since *B*, too, is an inertial observer, he is, by our definition, traveling at a constant velocity, and in equal intervals of time he will travel equal distances. Hence, all *B*'s positions on the space-time diagram will be found to lie on a straight line, and any other inertial observers likewise can be represented by straight lines.[1]

Fig. 13 is a space-time diagram of *A* and four other observers. Note especially the differences in the steepness (or slope, to use the conventional mathematical term) of the lines representing the observers *B, C, D,* and *E*. The steeper the line, the less does *x* change for a given change of *t* and hence the slower the motion (the less distance covered in a given time) of the observer relative to *A*. Thus *B* moves away from *A* fairly slowly and *C* rather faster; while *E*, whose *x*-coordinate is constant, is at rest relative to *A*. Both *B* and *C* move away from *A* since at later times (i.e., higher up on the diagram) their *x*-coordinates are greater. On the other hand, *D* approaches *A*. Light, being very fast,

[1] The reader must realize, however, that while space-time diagrams are extremely helpful in showing who is where and when, they are only *graphical representations* drawn by one observer. The times experienced or the distances measured by other observers *cannot* be found by measuring distances on the diagrams.

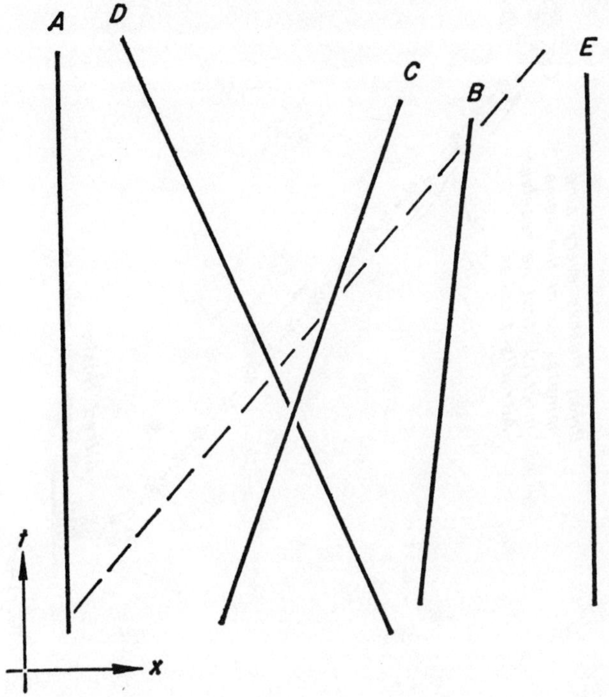

FIG. 13.

has a rather greater slope, and the dashed line is supposed to represent the journey of a flash of light.

Light is a wave phenomenon. In Chapter V we investigated another wave phenomenon, sound, and in particular examined the Doppler effect, the difference between the pitch of a wave when emitted and when received due to the relative motions of the source and the listener. In the case of sound, it will be remembered, it was necessary to know both the velocity of the transmitter relative to the air and the velocity of the receiver relative to the air. In the case of light there is no such thing as the air, and so all we have to know is the ve-

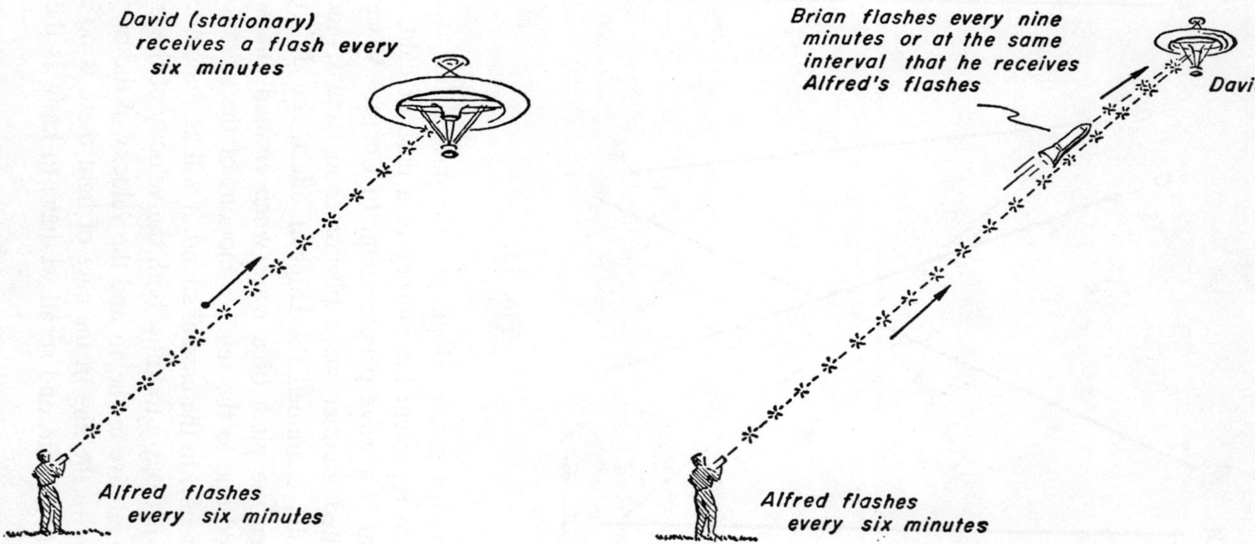
David (stationary) receives a flash every six minutes
Alfred flashes every six minutes
Brian flashes every nine minutes or at the same interval that he receives Alfred's flashes
David
Alfred flashes every six minutes

FIG. 14.

locity of transmitter relative to receiver, supposing both to be inertial and therefore moving with constant velocity.

Let us suppose that we have two observers a constant distance apart, and that the first observer, called Alfred, flashes an electric torch at regular intervals (Fig. 14). Purely in order to have a definite interval to talk about, and without any special significance attaching to this value, we suppose him to flash his torch every 6 minutes as measured by his watch. Then the light from each flash travels across to David, the second observer. Since the distance between them does not change in the course of time, it follows that each flash takes the same time to travel. Thus, if Alfred flashes his torch every 6 minutes by his watch, then these flashes will be received by David at equal intervals, also 6 minutes apart, by David's watch. We assume Alfred and David to be a good long distance apart. It might take each flash ten years (remember this is an idealized experiment) to reach David, but the significant thing is that the flashes, however long in transit, arrive at intervals of 6 minutes. There is nothing out of the way in this and the roles of the two observers could easily be interchanged.

Next, suppose we have a third observer, Brian, traveling quite fast from Alfred to David, who remain at rest relative to each other. (See Fig. 15. The reader is advised to open to this figure and leave it constantly available while following the argument.) Brian, while traveling from Alfred to David, also watches these flashes. As he moves away from Alfred every successive flash has to travel farther to reach him than the previous one, and is therefore longer in transit. Accordingly, by Brian's watch the flashes will not arrive every 6 minutes, but at longer intervals, simply because each flash

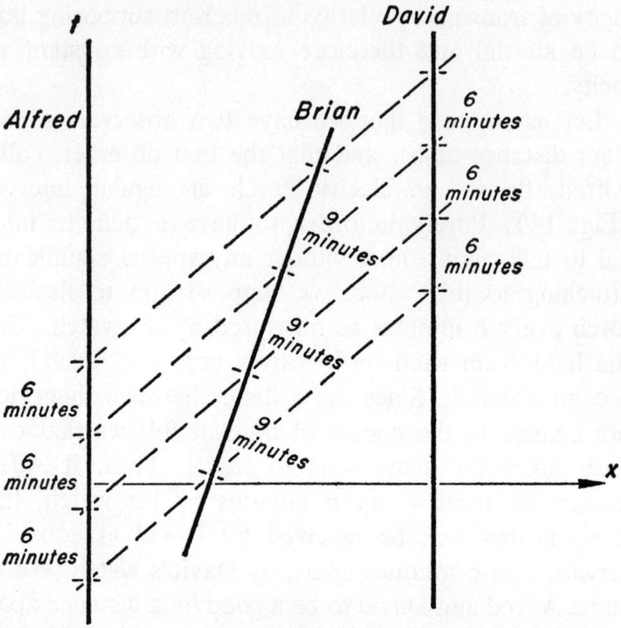

FIG. 15. *The ratio of intervals of reception to intervals of transmission is* 1 *between Alfred and David (at rest relative to each other);* $\frac{3}{2}$ *between Alfred and Brian, and* $\frac{2}{3}$ *between Brian and David.*

has a greater distance to cover than the preceding flash. For a suitable speed, which we need not work out now, we may suppose that the flashes are received every 9 minutes by Brian's watch. We always stress that every observer compares the arrival time of the flashes with his *own* watch, because otherwise we would have great difficulty in discussing just how long it takes him to see the other chap's watch.

Now suppose that Brian also has a light, a red one, which he flashes every time he sees a flash from Alfred. Since Brian sees these flashes every 9 minutes by his time, he will flash his red light every 9 minutes. The red flashes will then travel in company with the white flashes from Alfred, since they are emitted just as the flash of light from Alfred passes Brian. (We neglect the time it takes him to flash his torch.) These flashes from Brian, traveling in company with the flashes of Alfred's light, will be seen by David, who receives the flashes of Alfred (i.e., the white flashes) every 6 minutes. Thus he will receive the red flashes also every 6 minutes. Let us put this a little differently. Alfred and Brian are separating at a certain speed, which we have taken to be such that the interval between Alfred's flashes is multiplied by $\frac{3}{2}$ to give the interval between Brian's reception of the flashes (9 minutes instead of 6). Brian is traveling at the same speed *toward* David as he is receding *from* Alfred, but now it is a velocity of approach, not a velocity of recession, and the flashes emitted every 9 minutes are received every 6 minutes, reduced by a factor of $\frac{2}{3}$. Correspondingly, had Brian flashed his light every 6 minutes, the flashes would have arrived at David at 4-minute intervals. Thus, if instead of a velocity of recession one of approach is considered, the factor of $\frac{3}{2}$ between interval of transmission and interval of reception is changed to a factor of $\frac{2}{3}$. Our results are quite independent of whether any of these inertial observers can be regarded as at rest; indeed, the question of which of them is at rest does not make any sense. The only relevant matter is their *relative* velocity. What we have to remember is that if a velocity of recession leads to some ratio of transmitting interval to receiving interval, then the same velocity, but in approach, implies the reciprocal ratio.

In more general terms, if Alfred had flashed his light at intervals h, then David would have seen these flashes at intervals h, each flash taking the same time to reach him. Brian would have seen them at some interval kh by his watch so that k is the ratio of the interval of reception to the interval of transmission. If Brian flashes his torch at intervals kh by his watch, then these flashes, traveling in company with those emitted by Alfred, would be seen by David at intervals h, giving the reciprocal ratio $1/k$ between Brian and David.

The relation between any two of our inertial observers is completely specified by the ratio of the interval of reception to the interval of transmission. If this ratio is unity (as in the case of Alfred and David) then the two observers are at relative rest; if it is greater than one they are receding from each other; and if the ratio is less than one they are approaching each other. Note that the Principle of Relativity, by insisting on the equivalence of all inertial observers, makes it quite clear that the ratio must be the same whichever of a pair of inertial observers does the transmitting. It is through this rule that our work on light differs so sharply from the work on sound (Chapter V) where, it will be remembered, the speed of transmitter and receiver relative to the air had also to be taken into account.

A More Complicated Situation

We can now apply what we have found (importance of ratio of intervals–dependence of ratio only on pair of observers–reciprocal ratio for approach as compared with recession) to a more complicated situation. Let us consider another observer, Charles (David will no longer be required), whose velocity relative to Alfred is the same as Brian's, but in the opposite direction. (See Fig. 16 and Table II.) He, too, is therefore an inertial observer. We further suppose that Brian passes Alfred at 12 noon by Brian's watch, and that Charles

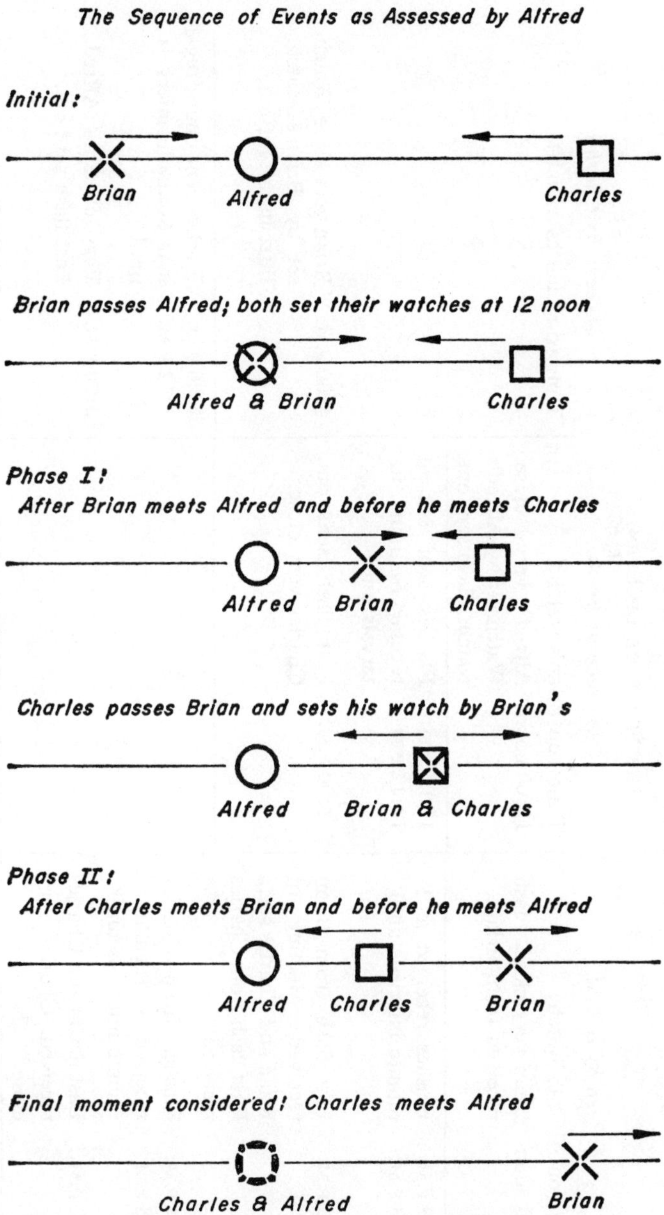
The Sequence of Events as Assessed by Alfred
Initial:
Brian
Alfred
Charles
Brian passes Alfred; both set their watches at 12 noon
Alfred & Brian
Charles
Phase I:
After Brian meets Alfred and before he meets Charles
Alfred
Brian
Charles
Charles passes Brian and sets his watch by Brian's
Alfred
Brian & Charles
Phase II:
After Charles meets Brian and before he meets Alfred
Alfred
Charles
Brian
Final moment considered: Charles meets Alfred
Charles & Alfred
Brian

FIG. 16. *The Sequence of Events*

TABLE. II

THE SEQUENCE OF EVENTS

As seen by Alfred		*As seen by Brian*		*As seen by Charles*	
Time (on Alfred's watch)		Time (on Brian's watch)		Time (on Charles' watch)	
12:00 noon	Brian very close. Flash received from Brian.	12:00 noon	Alfred very close. Watch set by Alfred's watch. Flash sent out.		
12:09 P.M. 12:18 P.M. ¦	Flashes received at 9-minute intervals until	12:06 P.M. 12:12 P.M. ¦	Flashes sent out at regular 6-minute intervals until		
1:30 P.M.	Last flash from Brian received. Meeting of Brian and Charles seen. First flash from Charles received.	1:00 P.M.	Last flash sent out. Charles very close.	1:00 P.M.	Brian very close. Watch set by Brian's watch. First flash sent out.
1:34 P.M. 1:38 P.M. ¦	Flashes from Charles received at regular 4-minute intervals until			1:06 P.M. 1:12 P.M. ¦	Flashes sent out at regular 6-minute intervals until
2:10 P.M.	Last flash from Charles received. Charles very close.			2:00 P.M.	Very close to Alfred. Last flash sent out.

passes Brian at 1 P.M. by Brian's watch. Some time later Charles passes Alfred. Since Charles and Brian have the same speed relative to Alfred and are in fact quite symmetrically placed, one hour will elapse according to Charles' watch between his meeting with Brian and his meeting with Alfred. The situation is that Alfred, Brian, and Charles remain in line through the experiment. Initially Brian is to the left of Alfred and Charles to the right of Alfred but much farther away. Brian and Charles are approaching Alfred at the same speed. Accordingly, first Brian passes Alfred, then Brian and Charles meet, and finally Charles passes Alfred; eventually Charles is to the left of Alfred, while Brian is to Alfred's right and much farther away. Both Brian and Charles are then receding from Alfred.

We focus our attention on the three meetings (first, Alfred and Brian; second, Brian and Charles; third, Charles and Alfred). As we have stated, the first two of these meetings are separated by one hour according to Brian's watch, he being the only observer present at both, and similarly the last two meetings are separated by one hour according to the watch of Charles, who is the only observer present at both of these meetings (Fig. 17). Suppose now that Charles adjusts his watch to read the same as Brian's when they are together. Thus it will read 1 P.M. then and hence it will read 2 P.M. when he meets Alfred. Suppose that Alfred also adjusted his watch by Brian's when they were together, so that it will have read 12 noon then. The question now is: What will Alfred's watch read when he and Charles meet? If, as we shall prove, it does not read 2 P.M., the time registered by Charles' watch, then we have established the route-dependence of time. It will be route-dependence in a slightly different form from that of mileage. It will depend on coming and going (i.e.,

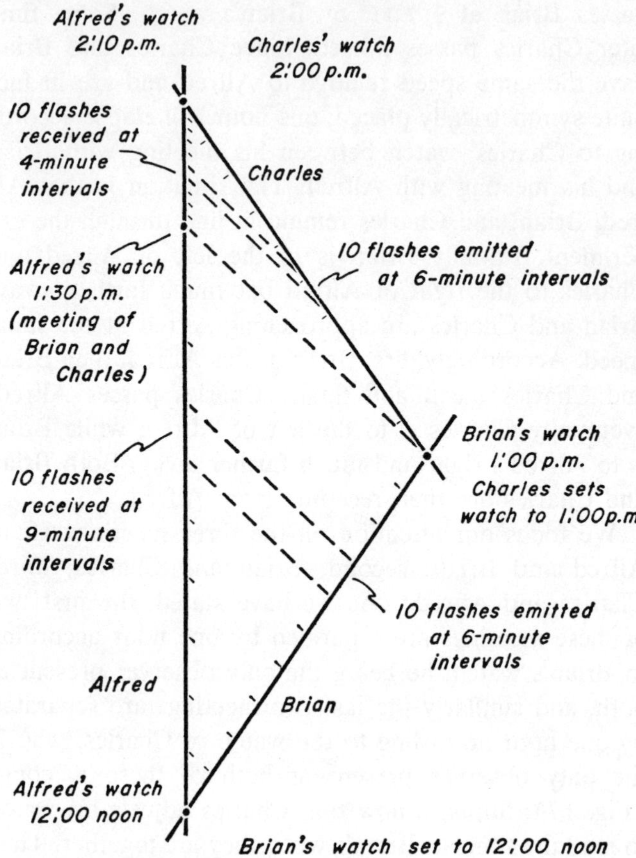

FIG. 17. *The Sequence of Events on a Space-Time Chart*

switching from one inertial observer to another) as opposed to staying with one and the same inertial observer.

To establish Alfred's timings, suppose next that Brian emits flashes every 6 minutes by his watch, the first one at his meeting with Alfred, the last one at his meeting with Charles. Between these two meetings, which occur one hour apart by Brian's watch, there are thus exactly ten intervals between flashes. Since Alfred and Brian are receding from each other with the same velocity as in the previous example, Alfred receives these flashes at 9-minute intervals by his watch. Alfred's watch read 12 noon when the first flash arrived since Brian was with him at the time and the transmission and reception of the signal were virtually simultaneous. Thus Alfred receives the last flash 90 minutes later–at 1:30 P.M. by his watch. This is the moment when light from the meeting of Brian and Charles reaches him–i.e., the moment when he "sees" them meet.

Next suppose that Charles, too, is sending out flashes at 6-minute intervals by his watch, the first when he meets Brian and the last when he is with Alfred. There are thus ten such intervals since the journey takes him one hour by his watch. By the previous example, these flashes are received by Alfred at 4-minute intervals, since Charles is approaching Alfred at just the speed at which Brian was approaching David. Thus these ten intervals occupy 40 minutes of Alfred's time. The first flash arrives at 1:30 P.M. by Alfred's watch since it was emitted when Charles met Brian, the light from this meeting arriving at 1:30 P.M. according to Alfred. Hence the last flash arrives 40 minutes later by Alfred's watch, which then reads 2:10 P.M. But this last flash was emitted by Charles when he was close to Alfred, at 2 P.M. by Charles' watch. Since the two observers are then so close together that light takes virtually no time to reach one from the other, the time of this meeting is 2 P.M. by Charles' watch and 2:10 P.M. by Alfred's.

To sum up: The first event we consider is the meeting of Alfred and Brian, at which both these observers set their watches to read 12 noon. At that moment

Brian sends out his first flash, which is received immediately by Alfred. During the next phase Brian is receding from Alfred, while Charles is approaching Alfred, though he is still beyond Brian. During this phase, which lasts one hour by Brian's watch, Brian emits flashes at 6-minute intervals, and Alfred receives them at 9-minute intervals (3:2 ratio). At the end of this phase Brian's watch reads 1 P.M., and Charles passes him, setting his watch by Brian's. Charles immediately starts to emit flashes at 6-minute intervals. Brian's last flash, emitted at this meeting, travels in company with Charles' first flash and they both arrive at Alfred at 1:30 P.M. according to Alfred's watch (ten intervals of 9 minutes length each after 12 noon). During the second phase Charles is approaching Alfred and sending out flashes at 6-minute intervals (Brian, beyond Charles, is no longer of interest). Charles' flashes are received by Alfred at 4-minute intervals (2:3 ratio). When Charles has traveled for one hour of his time since the meeting with Brian, he meets Alfred. Accordingly, by Charles' watch this meeting occurs at 2 P.M. The ten intervals between Charles' flashes occupy 40 minutes of Alfred's time. Since the first one arrived at 1:30 P.M. on Alfred's watch, the last one (emitted by Charles when with Alfred) arrives at 2:10 P.M., which is thus Alfred's time when Charles and he meet (Fig. 16 and Table II).

Relativity Explains a Supposed Discrepancy

Which of them is right? Of course, the answer is that they are both right; after all, if two motorists drive from New York to Boston and one clocks 230 miles and the other clocks 250 miles, we do not say that one of them was wrong; we merely say that one took a more direct route than the other. We need not suspect either milometer. What we have to get used to is that time, just like distance, is route-dependent. The time from the first meeting to the last via the meeting between Brian and Charles is shorter than the time from the first meeting to the last meeting, as measured by Alfred. It is not a question of the clocks of Brian or Charles having been "affected" by their speed. This would be as absurd a way of looking at it as to say that a motorist's milometer had been "affected" by his circuitous route in indicating a longer distance than some other motorist's has shown. It is not a question of there being anything wrong with the milometers or the watches; it is simply a fact that time is a route-dependent quantity, just as mileage is. What we have deduced is that, with the notions of relativity, we cannot maintain the idea that time is a route-independent quantity and hence that a public time exists. What does exist is private time and one that depends on the way one goes from one event to another one, whether with Alfred or with Brian and Charles.

Sometimes people have been confused by this result and complain, "But we can just as well regard Brian as standing still and Alfred as moving as Alfred standing still and Brian moving." This, of course, is perfectly true: the one thing that we cannot do is to regard both Brian and Charles as not moving. There is no way in which this problem is symmetric between Alfred, on the one hand, and Brian and Charles, on the other. There is one of Alfred but two of Brian and Charles, and there is no way of looking at it that can reduce more than one of the three to rest. One measurement is always carried out by Alfred himself; the other is a *combination* of measurements of Brian and Charles.

The Value of k: A Fundamental Ratio

It is clear that the discrepancy of ten minutes in two hours is highly velocity-dependent. We investigate the problem now in general terms. Suppose the ratio of the interval of reception to the interval of transmission between Brian and Alfred to be k. By what has been proved, it is then $1/k$ between Charles and Alfred.

We now suppose Alfred and Brian to set both their watches to read zero at their encounter, and we call Brian's watch reading zero + T at the moment when Charles shoots past him. Charles sets his watch by Brian's at that moment so that it also reads zero + T. Charles rushes toward Alfred at just the speed at which Brian is moving away from Alfred, and so Charles registers the same time T between passing Brian and Alfred as Brian did between meeting Alfred and Charles. Thus Charles' watch reads zero + T + T, or $2T$, when he encounters Alfred. (See Table III.)

Suppose now that Brian emitted a flash of light when he passed Alfred, and again when he passed Charles, i.e., at an interval T by Brian's watch. These two flashes will therefore be received by Alfred separated by interval kT. Since the first flash had no distance to travel, being emitted just when Brian was close to Alfred, it will arrive immediately (i.e., at Alfred's time zero) and so the second one will arrive at Alfred's time kT. Charles similarly emits flashes at both his encounters. They are separated by interval T by Charles' own watch, but since Charles is approaching Alfred, will be received at an interval T/k. The first of Charles' flashes is emitted at the same place and time as Brian's second flash (viz., at their encounter) and so travels with it, also

TABLE III

THE SEQUENCE OF EVENTS

As seen by Alfred	*As seen by Brian*	*As seen by Charles*
Alfred's time	Brian's time	Charles' time
0: Brian very close. Flash from him received.	0: Alfred very close. Watch set by his watch. Flash sent out.	
	T: Charles very close. Flash sent out.	*T:* Brian very close. Watch set by his watch. Flash sent out.
kT: Meeting of Brian and Charles seen. Flashes received from both.		
$(k+\frac{1}{k})T$: Charles very close. Flash from him received.		*2T:* Alfred very close. Flash sent out.

arriving at Alfred at time kT on Alfred's watch. Thus Charles' second flash reaches Alfred at time $(k+\frac{1}{k})T$ on Alfred's watch. Since it was emitted by Charles when he was close to Alfred, it arrives immediately so that $(k+\frac{1}{k})T$ is the reading of Alfred's watch when Charles passes him at a time $2T$ by Charles' watch. Thus the ratio of the time between Brian's and Charles' meetings with Alfred as measured by Alfred, to that measured by Brian and Charles, is

$$\frac{\left(k+\frac{1}{k}\right)T}{2T}=\frac{1}{2}\left(k+\frac{1}{k}\right)$$

This ratio depends very sensitively on the value of k as Table IV shows, which also gives the difference between the readings of Alfred's and Charles' watches for $T=1$ hour.

TABLE IV

k	$\frac{1}{2}\left(k+\frac{1}{k}\right)$	Time diff. $T=1$ hour
1	1	0
1.0001	1.000000005	1/30 millisec.
1.01	1.00005	1/3 sec.
1.25	1.025	3 min.
1.5	1.083	10 min.
2	1.25	30 min.
4	2.125	2 h. 15 min.
10	5.05	8 h. 6 min.
100	50.005	4 days 4 h. 36 sec.

While the connection between k value and velocity will only be found in the next chapter, it might be worth mentioning here that $k=0.0001$ corresponds to a speed of 19 miles per second, which is the orbital speed of the Earth and about four times the speed of an orbiting satellite, while $k=10$ corresponds to 90 percent of the speed of light and $k=100$ to 99.98 percent of that speed.

In our daily lives the ratio k is always so close to unity that the discrepancy between the clock readings is wholly inappreciable, and thus the illusion of a route-independent time is fostered.

Now suppose that Brian traveled with his young son, whom he had entrusted with a watch, and that when Charles rushed past, Brian threw the boy across to Charles, who caught him nicely, and then traveled with the boy to his meeting with Alfred. Then you might argue, "But is not there as much right for Alfred to have regarded himself as at rest and the boy as a traveler, as for the boy to have regarded himself at rest and Alfred as the traveler?" But this is unsound: Brian, Charles, and Alfred are all inertial observers; they do not suffer any jerks; if any of them had carried a bag of raw eggs at the beginning of our experiment, the bag would still be in order at the end. But the boy is not an inertial observer; he changes his speed very suddenly and sharply, receiving a severe jerk. Had he been entrusted with a bag of raw eggs they would have been in a horrible mess at the end. There is no comparison between Alfred, who is inertial, and the boy, who is not. Thus there is no symmetry and no difficulty arises in this manner. If we do not suppose that the shock of being thrown from Brian to Charles was too much for the boy or his watch, a question that we shall investigate in more detail in a later chapter, then the boy at the final meeting will be two hours older than at the initial meeting, whereas Alfred will have aged two hours and ten minutes. Again we can compare this with the mileages. Of all the mileages between two towns, the shortest one is obtained by traveling in a straight line; all others are longer. With times in relativity the opposite holds; between any two events the time clocked by an inertial observer is the maximum, and the time clocked by any other is shorter. Thus, traveling could keep one young, were it not for the possibly disastrous experience of being thrown from Brian to Charles. The relative speeds of the observers in this example will be evaluated in the next chapter. It is clear that they are very large, many tens of thousands of miles per second. Such speeds are so far outside the experience of everyday life that it is not in the least surprising that the results of such strange experiences should be unfamiliar.

CHAPTER IX

VELOCITY

In the last two chapters we deduced a variety of results for inertial observers, given the ratio of the interval of reception of two flashes of light by one observer to the interval of transmission by the other. How can we relate this ratio, so fundamental to our working, to the relative velocity, which has a far more direct appeal?

Let us go back to Alfred and Brian and their relative motions of the previous chapter. For the sake of realism, we shall equip them with radar instead of lights (the principle is the same) and assume that Brian, on receiving a signal from Alfred, can respond with his radar instantaneously. The situation then, you will note, will not be unlike sending out a radar pulse and receiving its reflected echo back from the Moon, say, or from some artificial satellite whose position we wished to measure accurately (Fig. 18).

Since the given ratio of the intervals of reception and transmission is 3:2, Alfred's signals, emitted at 6-minute intervals, are received by Brian at 9-minute intervals. It is supposed also that the watches of both indicate 12 noon when the two observers pass each other. Then, if Alfred sends out one pulse at 12 noon and a second pulse at 12:40 P.M., the two signals will arrive at Brian sixty minutes apart—in Brian's reckoning (Fig. 19). Since Brian is with Alfred at 12 noon on Alfred's watch, he receives Alfred's first signal at that hour (12

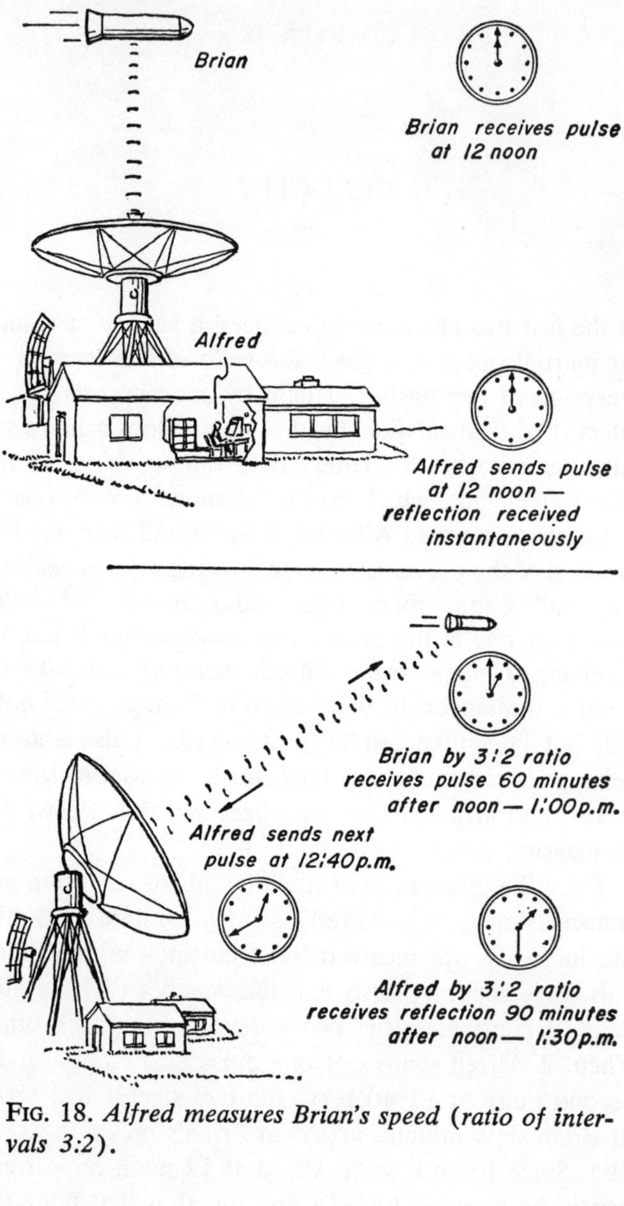
Brian
Brian receives pulse at 12 noon
Alfred
Alfred sends pulse at 12 noon — reflection received instantaneously
Brian by 3:2 ratio receives pulse 60 minutes after noon— 1:00p.m.
Alfred sends next pulse at 12:40p.m.
Alfred by 3:2 ratio receives reflection 90 minutes after noon— 1:30p.m.

FIG. 18. *Alfred measures Brian's speed (ratio of intervals 3:2).*

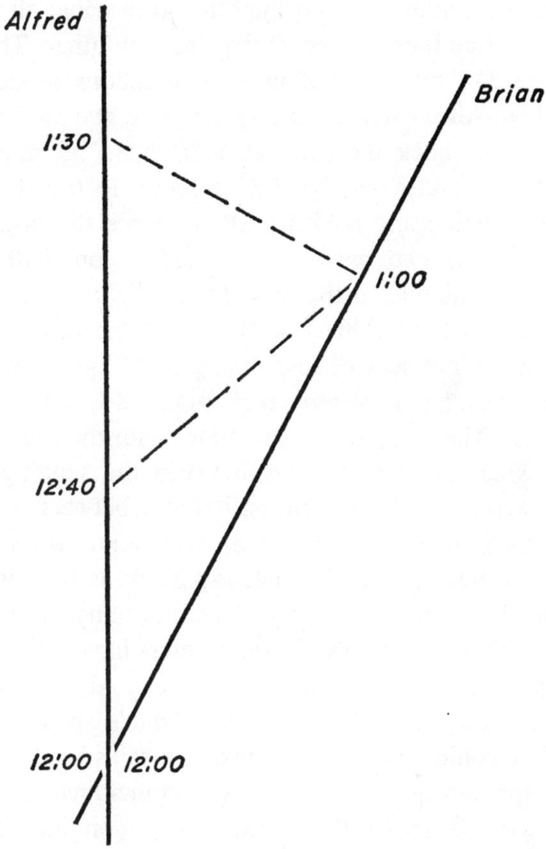

FIG. 19. *Space-Time Chart of Alfred Measuring Brian's Speed*

noon by his watch also) and the second one at 1 P.M., which is the moment of his meeting with Charles. Brian replies instantaneously with a pulse and this response, sent out 60 minutes after noon by his own watch, is by the 3/2 ratio received by Alfred at 90 minutes after noon (Fig. 19). Remember (from the discussions on pages 78–79) that neither Alfred nor Brian can say that he is standing still and the other is moving; all they can say is that they are receding from each other. Therefore, the 3/2 ratio applies in both directions of signaling. Thus, Alfred sends out a radar pulse at 12:40 P.M. and receives back the echo at 1:30 P.M., 50 minutes later. It has taken the signal 50 minutes to travel from Alfred to Brian and back to Alfred, twice the distance separating the two observers. Therefore, one-half this time, or 25 minutes, is the time it takes light (or a radar pulse) to go from Alfred to Brian. So the distance of Brian at the moment of responding is 25 light minutes.

But how long on Alfred's reckoning did Brian take to get there? The time Alfred associates with the reflection of his radar pulse is halfway between the sending out and the receiving back—that is, halfway between 12:40 P.M. and 1:30 P.M., which is at 1:05 P.M. Alfred has no choice but to take this halfway mark as the instant of reflection. For the velocity of light is unity by definition and thus in Alfred's view light must have taken just as long to get to Brian as to get back. Above all, he must not attempt to "correct" for Brian's speed, since the flash could have been returned just as well by an object moving quite differently but coinciding momentarily with Brian at the instant of responding. Thus Alfred cannot help assigning the time 1:05 P.M. to this instant. He therefore arrives at the answer that Brian took 65 minutes (from 12:00 to 1:05) to cover a distance that light covers in 25 minutes. Accordingly, Brian's speed relative to him is $\frac{25}{65} = \frac{5}{13}$ of the velocity of light in Alfred's reckoning. With the conventional value for the velocity of light, this fraction gives the speed of Brian relative to Alfred as 71,700 miles per second—a very respectable speed by our standards, but yet a perfectly possible speed. By virtue of our assumptions, the speed of Charles relative to Alfred is precisely the same.

EINSTEIN'S LONG TRAINS

Einstein published his Special Theory of Relativity, which is the subject of our discussion here, in 1905. The airplane had been off the ground for only two years; the railroad train was the ultimate in speed, and it seemed highly unlikely that trains would ever travel at more than 150 miles an hour. When Einstein in 1916 wrote a book on Relativity for the general public, he could think of no better example to illustrate his ideas than to imagine indefinitely long trains running past indefinitely long embankments at speeds approaching the velocity of light! For more than forty years his followers in the attempt to explain Relativity were in the same fix; even the 1958 revision of Bertrand Russell's *The ABC of Relativity* considers various problems in relation to a train running at a speed three-fifths the velocity of light along a straight track of indefinite length. The writers had no choice. Farfetched as they were, those trains afforded about the only possible images that would fall within the layman's understanding and not be dismissed as Jules Verne absurdities. Small wonder that the public should have regarded Relativity as at best the impractical speculation of philosophers, or at worst as learned phantasy.

Today all this has changed. We send rockets to the Moon and to the vicinity of Venus. The most stubborn skeptic no longer doubts that space stations of some sort will exist within the lifetime of the youngest readers of these pages. Russian and American astronauts circle the Earth at speeds approaching 20,000 miles an hour, and while our Brian's 71,700 miles per second is a far stretch from that figure, yet we can think realistically of speeds quite beyond the ken of our fathers and grandfathers. Every day experimenters at the great accelerating machines (the "atom-smashers") work with speeds well over nine-tenths the velocity of light—relativistic effects are the regular order of their business. In colour TV sets, the velocity of the electrons is high enough for the designer to have to take relativistic effects into account! In a very few years, Special Relativity has come down from the clouds of phantasy or philosophic speculation to its rightful foot-hold on the solid ground of the public domain.

It is in the nature of the human mind that learning is easier when a demonstrable *need* to learn exists. Our fathers had no actual need to understand Relativity, but we have, and we can address ourselves to the adventures of Alfred, Brian, and Charles without the emotional misgivings that, forty years ago, upset the passengers on Einstein's indefinitely long trains. Alfred, Brian, and Charles are no less fictional, but their maneuverings in space are representative of situations which, in more complex detail and refined form, command the attention and challenge the laboratory skills of today's scientists and engineers. So, bearing in mind always that it is a grasp of the concepts underlying quite real phenomena we are seeking, let us return to our three inertial observers' investigations of time differences in Special Relativity.

Determining Relative Velocities by the Radar Method

Let us go back to the motions of our observers in Chapter VIII and consider a slightly different situation. Suppose again that Brian and Charles have the same speed relative to Alfred and that one hour elapses by Brian's reckoning between his meetings with Alfred and Charles, and correspondingly one hour by Charles' watch between Charles' meetings with Brian and Alfred. Now, however, we suppose that there is a factor of three instead of three-halves between the length of the intervals at which Brian emits flashes and the length of the intervals at which Alfred receives them (Fig. 20).

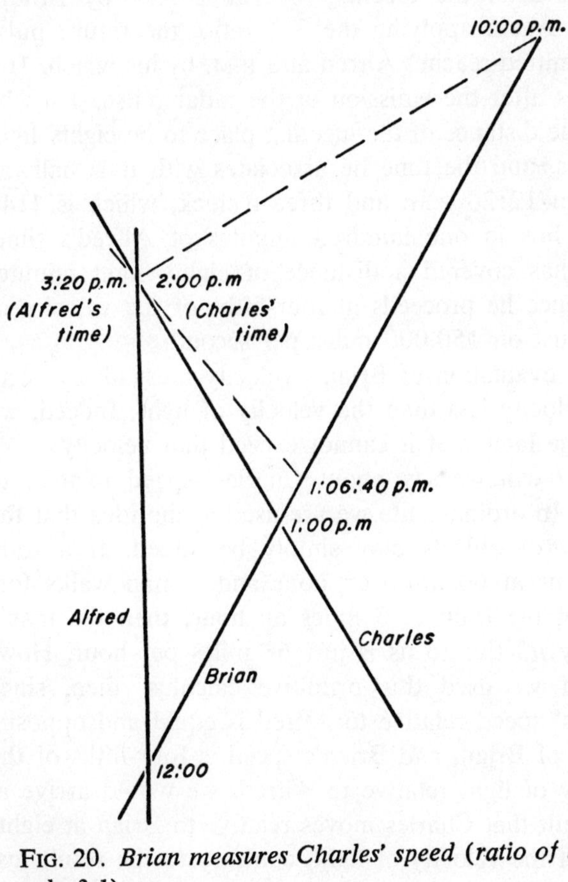

FIG. 20. *Brian measures Charles' speed (ratio of intervals 3:1).*

Then Brian's one hour between his meetings turns into three hours for Alfred, and Charles' one hour into one-third of one hour, that is, twenty minutes; so, as stated previously, three hours and twenty minutes emerge as the time on Alfred's watch between his meetings with Brian and with Charles. If Alfred emits a radar pulse at 12:20 P.M. (twenty minutes after Brian passed him), then by the 3/1 ratio this pulse reaches Brian sixty minutes after the meeting–i.e., at 1 P.M. by Brian's watch. Again applying the 3/1 ratio, the return pulse then emitted reaches Alfred at 3 P.M. by his watch, 160 minutes after the emission of the radar pulse. Thus he finds the distance of the meeting place to be eighty light minutes, and the time he associates with it is halfway between 12:20 P.M. and three o'clock, which is 1:40 P.M. Thus in one hundred minutes of Alfred's time, Brian has covered a distance of eighty light minutes and hence he proceeds at four-fifths of the velocity of light–just on 150,000 miles per second.

This evaluation of Brian's velocity must always lead to a velocity less than the velocity of light. Indeed, we shall see later that it cannot exceed that velocity.

What can we say about Charles' speed relative to Brian? In ordinary life we are used to the idea that the velocity of objects may simply be added. If a train passes us at 60 miles an hour and a man walks forward in the train at 3 miles an hour, then the man's velocity relative to us is just 63 miles per hour. However, if we used this primitive calculus, then, since Charles' speed relative to Alfred is equal and opposite to that of Brian, and Brian's speed is four-fifths of the velocity of light relative to Alfred, we would arrive at the result that Charles moves relative to Brian at eight-fifths of the velocity of light, contrary to the result just stated.

But we should not jump to such conclusions. We have given a perfectly good method for working out velocities just now using precisely the radar type of method that is actually often employed. Can we not use the same method directly to establish Charles' speed relative to Brian's in this example? Suppose Brian wishes to determine the distance of the meeting of Alfred and Charles from himself. To use radar, he has to send out a pulse of light a little while after Charles has left him, because light travels faster than Charles does, and then he must wait until the reflection from that meeting reaches him. In the fast motion just considered, Charles and Alfred met 200 minutes, by Alfred's clock, after Brian left Alfred. When does Brian have to send out a radar pulse to arrive at Alfred at that time? Since the ratio of the interval of transmission and the interval of reception is three now, it follows that by Brian's watch he should have sent out that pulse after an interval of one-third of 200 minutes–that is, 1 hour, 6 minutes, and 40 seconds after he left Alfred (Fig. 20), or 6 minutes and 40 seconds after Charles flashed past him. The radar pulse will come back to Brian at three times the interval by Alfred's clock between Alfred's meetings with Brian and Charles. Three times 3 hours and 20 minutes is 10 hours, and thus Brian receives the return pulse at 10 P.M., 8 hours, 53 minutes, and 20 seconds after the pulse was emitted. Hence, the distance of the meeting of Alfred and Charles in Brian's measurements is 4 light-hours, 26 light-minutes, and 40 light-seconds. The time Brian associates with this meeting is halfway between the sending out and the reception of the ray, which is 4 hours, 33 minutes, and 20 seconds after Charles left him. Accordingly, in Brian's reckoning, Charles has covered 4 light-hours, 26 light-minutes, and 40 light-seconds in

4 hours, 33 minutes, and 20 seconds; a velocity close to the velocity of light, in fact, about 97.5 percent of it, but not exceeding it. Thus by adding two velocities, each 80 percent of the velocity of light, we obtain only 97.5 percent of the velocity of light.

If this argument is pursued, we find that we can add any number of velocities less than the velocity of light but never get a velocity equal to or exceeding the velocity of light; we always stay below it. Of course, this means that adding large velocities is not quite as simple as adding small velocities, but it is a direct consequence of defining in a sensible fashion the method of measuring and determining velocities. Thus the velocity of light appears in this context like a rainbow; try as we will to reach it, we never can.

Looked at a little differently, this result is really an obvious consequence of our assumptions. If one observer moves more slowly than light, then a light ray emitted by him gets everywhere before he does, and every other observer will agree on this. To every observer, however moving, the emitter will appear to move more slowly than light. Since the velocity of light is the same for all observers, his velocity relative to anybody will be less than the velocity of light.

The Relationship between k and v

To work this through in general, suppose that the ratio of intervals is k, and that Alfred sends out his radar pulse at time T after he and Brian pass each other, a moment which for convenience is counted as the common starting point of both their time reckonings. Then this pulse will be received by Brian at time kT, and Brian's response will reach Alfred at time $k \times kT = k^2T$. Thus Alfred's time interval between transmission and reception is $(k^2-1)T$ and so the distance of Brian from Alfred at the moment of responding is $\frac{1}{2}(k^2-1)T$. Again, since the velocity of light is the same (unity) in each direction, the moment of response is counted by Alfred as occurring halfway between transmission and reception–i.e., at time $\frac{1}{2}(k^2+1)T$. Between zero time and this instant Brian has changed his position from proximity to Alfred to distance $\frac{1}{2}(k^2-1)T$. Accordingly Brian's velocity v is the ratio of these quantities, so that

$$v=\frac{k^2-1}{k^2+1} \qquad (1)$$

Note that v is a pure number which is a natural consequence of making the velocity of light unity. It will also be observed that $k=1$ (equal intervals) corresponds to the state of relative rest $v=0$, that replacement of k by $1/k$ simply changes the sign of v, in accordance with the results of Chapter VIII, and, finally, that for all values of k the velocity v will be between -1 (velocity of light in approach) and $+1$ (velocity of light in recession). This is a clear consequence of the method of measurement for $v=+1$, since otherwise the radar pulse could not catch Brian, and of the replacement of k by its reciprocal for -1. Solving equation (1) for k we have

$$k=\left(\frac{1+v}{1-v}\right)^{1/2} \qquad (2)$$

The relation between v and k is given in the table below, including conventional values of v in miles per second.

k	1	1.001	1.1	1.5	3	10	100
v	0	0.001	0.095	0.385	0.8	0.98	0.9998
v (*miles per second*)	0	186	17,700	71,700	149,000	182,000	186,000

Velocity Composition

Next consider the composition of velocities. Consider three observers, Alfred, Brian, and Edgar, such that the ratio of intervals of transmission to intervals of reception is k between Alfred and Brian and k^1 between Brian and Edgar, and suppose that, seen from Alfred, Edgar is beyond Brian (Fig. 21). Signals emitted by Alfred at interval T are received by Brian at interval kT. If Brian sends out flashes whenever he re-

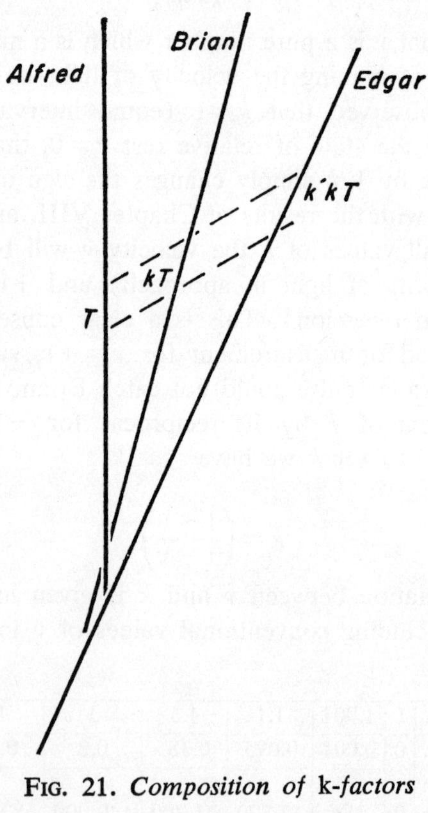

Fig. 21. *Composition of* k*-factors*

ceives signals from Alfred (i.e., at intervals kT), Edgar will receive them at intervals kk^1T, simultaneously with Alfred's signals. Thus the ratio between Alfred and Edgar is kk^1, or, in other words, *k values multiply*. This is the fundamental rule of velocity composition, which at low speed reduces to the familiar direct addition of velocities. Note that whatever number may result from the multiplication of k values, the velocity corresponding to the final k value must still be less than the speed of light by (1).

To work out the effect of k multiplication on velocities, let v be the velocity Alfred-Brian, v^1 the velocity Brian-Edgar, and w the velocity Alfred-Edgar. Then

$$k = \left(\frac{1+v}{1-v}\right)^{1/2}, \quad k^1 = \left(\frac{1+v^1}{1-v^1}\right)^{1/2}, \quad w = \frac{k^2(k^1)^2 - 1}{k^2(k^1)^2 + 1} = \tag{4}$$

$$= \frac{\left(\frac{1+v}{1-v}\right)\left(\frac{1+v^1}{1-v^1}\right) - 1}{\left(\frac{1+v}{1-v}\right)\left(\frac{1+v^1}{1-v^1}\right) + 1} = \frac{v+v^1}{1+vv^1}$$

One sees readily how for small v and v^1 their sum equals w, while for any v and v^1 not exceeding unity w will also not exceed unity. All this confirms the results previously obtained in special cases.

Proper Speed

There is another way of introducing velocity into the theory, which is not as close to our usual notion of velocity as distance covered per unit time, but is a useful quantity.

Alfred, by using his radar technique, employs virtually the only possible method for finding the distance of the meeting between Brian and Charles, but in order to determine the time elapsed he had to work out the mean of the time of emission and time of reception of his radar pulse. Instead of doing that, he could have looked at Brian's watch and formed the ratio of the distance Brian covered in Alfred's reckoning to the time Brian took to cover this distance in *Brian's own reckoning*. We have, therefore, a somewhat hybrid quantity, which is called proper speed. The word proper is introduced because we divide by the time that belongs to Brian—Brian's proper time. In Brian's reckoning, 60 minutes elapsed between his leaving Alfred and meeting Charles. He thus covered 25 light-minutes of his own time, and thus his proper speed is $\frac{5}{12}$ instead of the $\frac{5}{13}$ obtained previously for his ordinary velocity. If we go over to the faster motion that we considered next, with three-to-one ratio of intervals of transmission and reception between Brian and Alfred, then, as will be recollected, Brian's distance from Alfred when he met Charles was 80 light-minutes. Again, in Brian's reckoning 1 hour has elapsed since he left Alfred, and so his proper speed in this case is $\frac{4}{3}$. Thus the proper speed of a body can be greater than unity. Indeed, it can grow without limit, and the proper speed of light, on this basis, is infinite, a result that can readily be deduced from what has been said earlier.

For many purposes in calculations in the Theory of Relativity, proper speed is easier to work with than the velocity. The point is that in an ordinary velocity we divide a distance, which different observers will measure differently, by time, which different observers will measure differently. On the other hand, in the proper speed, we still have the distance which will be measured differently by different observers, but at least the time is the proper time measured by the observer himself and therefore is something on which everybody agrees. Brian's own clock readings between two events in Brian's life will look the same from wherever they are viewed.

The Unique Character of Light

A further result can be deduced from our considerations. We have seen earlier that as we increase the ratio of the interval of reception to the interval of transmission between Brian and Alfred from three-halves to three, the discrepancy in the time measured between Brian's meeting with Alfred and Charles' meeting with Alfred, as measured either by Alfred or by Brian and Charles, increases. We have taken it that in both cases the Brian/Charles measurement of time was 2 hours, but the Alfred measurement of time went up from 2 hours and 10 minutes to 3 hours and 20 minutes. Clearly, the larger we take the ratio, the longer Alfred's time will become. Conversely, we could keep Alfred's time the same by diminishing the timings of Brian and Charles as we increase the ratio. As the velocity of Brian and Charles viewed by Alfred gets nearer and nearer the velocity of light, light finds it harder and harder to catch up with them, and so the ratio of interval of reception to interval of transmission increases from three-halves to three, and on and on without limit as we consider higher and higher velocities. In order, then, to get a fixed time for Alfred between his meeting with Brian and his meeting with Charles, we have to cut down the time taken from the first meeting to the last meeting as measured by Brian and Charles. If we go to the limit and have Brian and Charles actually riding on light waves, no time will have elapsed by their reckoning.

We cannot imagine real people traveling at the speed of light, because, as we have seen earlier, there is no way of getting them up to that speed, but we can think of light itself. We could think of a mirror at the meeting of Brian and Charles, from which the light emitted bounces back and eventually reaches Alfred. If, then, this light ray carried a clock (rather an absurd idea but one which we can approach in the limit) then this clock would have registered no time whatever between leaving Alfred and coming back to Alfred, adding the time from Alfred to the mirror and back from the mirror to Alfred. We can put this differently; we can simply say that light does not age; there is no passage of time for light. This view helps to make the unique and universal character of light somewhat clearer. It cannot change once it has been produced, owing to the fact that it does not age, and therefore it must remain the same.

CHAPTER X

COORDINATES AND THE LORENTZ TRANSFORMATION

So far we have derived all the consequences of Einstein's Principle of Relativity by using the factor *k*. We have come to new insights concerning the nature of time and indeed this *k* calculus can be used a great deal further, in order to develop in particular those consequences of the principle that can be observed readily and therefore afford the most powerful support for the principle. On the other hand, the textbooks published hitherto have all worked with coordinates and transformations of coordinates. Of course, this treatment is completely equivalent to ours. But it is of some advantage to make a connection with this more usual mathematical derivation. As a result, this chapter will altogether be a little more mathematical than either its predecessors or the later ones, but it is hoped that it will particularly benefit the reader who has had some previous knowledge of the theory.

The Meaning of Coordinates

The mathematician uses coordinates to fix the position of a point, and so it may be of advantage to discuss the whole meaning of coordinates. The simplest case arises when one works in a plane, say on a sheet of paper, on a blackboard, or on the floor of a room. Then one can specify the position of any point by giving its perpendicular distances from two lines at right angles to each other—the two axes of coordinates. In the usual language, one coordinate is called x, and the other coordinate is called y; the line on which y vanishes is then called the x-axis, and, similarly, the line on which x vanishes is called the y-axis. Given the two axes and given any pair of numbers x, y, one can readily find the point in the plane corresponding to this pair of numbers and, conversely, given any point in the plane, by simply measuring its perpendicular distances from the two axes, one can find its coordinates. This is therefore a very simple and ready method for specifying the position of points in a plane and is, indeed, used a great deal.

The one difficulty, but an unavoidable one, is that the choice of axes of coordinates is arbitrary. We can work with one pair of axes just as well as with any other pair of axes. What happens when we change the axes of coordinates? There are two rather different ways to do it. In one, which is almost trivial, we simply displace the axes, but in parallel; that is to say, the orientation of the new axes is just the same as the orientation of the old axes—they just meet in a different point. If we call the new coordinates of a point x' and y', then it is clear that these new coordinates will be related to the old coordinates by the simple addition of numbers. We get the name of any point in the new system by taking its name in the old system and simply adding one number to x and another number to y. We use the same numbers for this addition wherever the point may be because these numbers are in fact simply the coordinates of the old origin in the new coordinates.

A much more interesting transformation, and one of great importance for our work, arises when instead of displacing the axes parallelly we turn our pair of axes through an angle. Even if at first we keep to the same origin and merely rotate the system of axes, then it is clear that the transformation between the old coordinates and the new ones will be a little complicated. We need not discuss it in any detail at all, but what will be clear—and this is the crucial point—is that the new x-coordinate will depend both on the old x- and on the old y-coordinate, and similarly for the new y-coordinate. In other words, the old coordinates get jumbled up when we wish to calculate the new ones.

This complication is quite familiar from everyday life. When we look at a house we call one side its width, and the other its length, and if we go around the corner and look at it from elsewhere, we might just as well refer now to what was previously the width as the length, and vice versa; and if we look at a house with a really complicated ground plan askew, then the dimensions may get jumbled up—the new width may be some combination of old width and the old length. It is because of this that the mathematician refers to a plane as having two dimensions. One needs *two* numbers to specify a point in a plane and these two numbers get mixed up with each other when we rotate our axes. This is the important distinction between the coordinates and other properties. For example, if we were designing, shall we say, the under-floor heating system of a house, we might be greatly interested in the temperature of each point of the floor, and we might say that the temperature counted just as much in its specification as its distance from the two walls; but although we would now have three numbers for each point of the floor, namely, its distance from the two walls and its temperature, we would not call the temperature the third dimension; and we would not call it the third dimension for the very good reason that there would be no transformation of coordinates that would be in the least sensible and have any meaning in which the temperature got mixed up with the other two coordinates.

This being able to "get mixed up with each other" is the crucial point about dimensions. In a general transformation of coordinates we both shift the origin and rotate the axes of coordinates, and if the point were represented by a pair of numbers x, y, in the old system of coordinates, then it will be represented by quite a different pair of numbers x', y', in the new system of coordinates. But there is, nevertheless, a subtle connection between the two different systems of coordinates. Let us suppose we have two points, x, y, and $\bar{x}$, $\bar{y}$; then in the new system of coordinates the first point will have coordinates x', y'; the second one $\bar{x}'$, $\bar{y}'$; but the one thing that must come out just the same in the one system of coordinates as in the other is the distance between the two points. In other words, the four numbers that represent the two points must possess a combination–namely, the distance between the two points–that is just the same in the old system of coordinates as in the new system of coordinates. Such a quantity is something that the mathematician calls an invariant, because it does not change as he alters the system of coordinates.

Rotation of Axes

Since we regard the transformation of just shifting the axes in parallel as trivial, we might as well shift the axes of coordinates until the origin is in one of the two points, say the point $\bar{x}$, $\bar{y}$, and then the change of system of coordinates is just a rotation of the axes about the origin. What is the distance, then, of the point x, y, from the origin? A simple application of Pythagoras' theorem shows that the square of the distance is simply $x^2 + y^2$.

If we rotate the axes about the origin, then the distance of the point from the origin must be the same as before, and so we have the theorem that in a rotation of coordinates the expression $x^2 + y^2$ is transformed into itself. It does not change its value in the least. So the upshot of all this is that we call a plane two-dimensional because, first of all, we need two coordinates to specify a point in the plane; secondly, because these two coordinates get jumbled up when we change our axes of coordinates; and, thirdly, because there exists an invariant, a combination of the coordinates, that does not change when we change the coordinates. As an example, suppose that the (x', y') axes are inclined at 30 degrees to the (x, y) axes. A little triangulation (Fig. 22) then shows that

$$x' = \frac{\sqrt{3}}{2}x + \frac{1}{2}y$$

$$y' = -\frac{1}{2}x + \frac{\sqrt{3}}{2}y$$

Thus the invariant

$$\begin{aligned} x'^2 + y'^2 &= \left(\frac{\sqrt{3}x + y}{2}\right)^2 + \left(\frac{-x + \sqrt{3}y}{2}\right)^2 \\ &= \frac{4x^2 + 4y^2}{4} \\ &= x^2 + y^2 \end{aligned}$$

The space in which we live is called three-dimensional because we need three coordinates, length, width, and height, or x, y, and z, to specify the position of a point. We make this specification in just the same manner as in the case of the two dimensions. We take three axes, our axes of coordinates, at right angles to each other, and then the perpendicular distance of a point

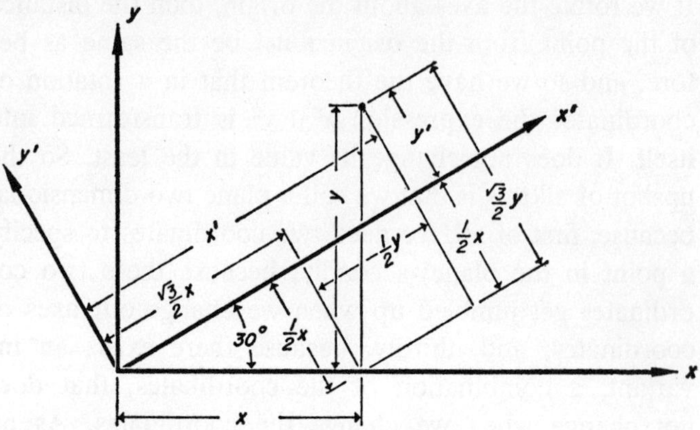

FIG. 22. *The Rotation of Axes*

from the plane containing the axes of x and y is its z-coordinate, and so on.

Again, we have the situation that there are three numbers that we need to specify a point, that these numbers get jumbled up when we change coordinates, and that there is an invariant—namely, the distance between points—which does not change when coordinates are changed. Once again we have two changes of coordinates, an almost trivial one in which we just shift the origin, the shift adding constants to the coordinates of all points, and the much more complicated change in which we rotate the coordinates so that the new ones are askew to the old ones. If we so rotate the coordinates, keeping the origin fixed, then again applying Pythagoras' theorem, we find that now $x^2+y^2+z^2$ remains unchanged. This quantity, in fact, is the only thing that is unchanged, the only invariant if we allow ourselves to turn the coordinates in a completely general manner.

It is useful, however, to think about this transformation in a little more detail. Suppose that somebody always draws his z-coordinate vertically upward; then the only change of coordinates that he allows himself is a rotation of the axes of x and y in the horizontal plane. It does not matter how much he changes these—he will always find the same z-coordinate as before. In the view of such a person, therefore, there are *two* invariants in space, the z-coordinate itself and $x^2 + y^2$, which remains unchanged when we turn the axes in the plane. This will be reasonable as long as he works only in a sufficiently small region for the vertical to have a perfectly clear meaning, and to be the same everywhere. But if he wants his coordinates to span a continent, then of course the vertical direction at one point is not the same as the vertical direction at another; and there is no particular reason why one of these should always be considered to be the vertical, the z-coordinate. He will then, therefore, as soon as his horizon is wider, consider more transformations of coordinates, even those in which the z-axis becomes inclined, and when this happens he finds that instead of his two invariants, z and $x^2 + y^2$, he has, in fact, only one invariant, $x^2 + y^2 + z^2$. Before he has gained this insight he might well have said, "Height!—that is something entirely different from width and length; there is no way in which the two ever get mixed up and they are completely different kinds of quantities." But after he has learned to consider inclined coordinates, he says, "Of course, x, y, and z are all the same; after all, they get all mixed up when I tilt my coordinates and they all enter into the single invariant there is."

THE LORENTZ TRANSFORMATION

How does all this fit into Relativity? It will be recalled that our guiding principle throughout was Einstein's

Principle of Relativity—that all inertial observers are equivalent. It was this principle that enabled us to make far-reaching deductions about what occurs at high speeds—that is, at speeds comparable to the velocity of light. The most important insight that we gained was that time is private rather than public, that the time measurements of different inertial observers do not necessarily fit together. This was the most striking result and very much in contrast with what had previously been thought. At low speeds all times fit together forming a public time, and so anybody used to working at low speeds only would have got the illusion that time was an invariant, in just the same way that somebody who had worked only in a small region of the earth, with the vertical clearly distinguished, might have thought that the z-coordinate was an invariant. We have already shown in Chapter VIII that this is not so, that the time does change when we consider high speeds; that therefore time cannot be regarded as an invariant. How does time transform as we go from one inertial observer to another one? We can now use the k calculus to establish these transformations, which are called collectively the *Lorentz*[1] *Transformation.*

Let Alfred use coordinates t, x and Brian coordinates

[1] Lorentz developed his Transformation in the course of a mathematical study of electromagnetism and applied it to an attempt to explain the Michelson-Morley experiment. G. F. Fitzgerald, an Irish physicist, had argued that the negative result of the experiment could be expected if the length of moving bodies, as measured by a stationary observer, underwent a contraction in the direction of motion. The Lorentz Transformation fits the hypothesis exactly. Einstein developed the same equation, but from a different argument. It has been customary to follow Einstein's derivation and to develop Special Relativity from the equation relating the space and time coordinates of two observers in uniform motion relative to each other.

t^1, x^1, so that Alfred is $x=0$, Brian is $x^1=0$, and at the meeting of Alfred and Brian $t=t^1=0$. Consider an event which, seen by Alfred, is beyond Brian (Fig. 23). Alfred emits a radar pulse at time $t-x$ and re-

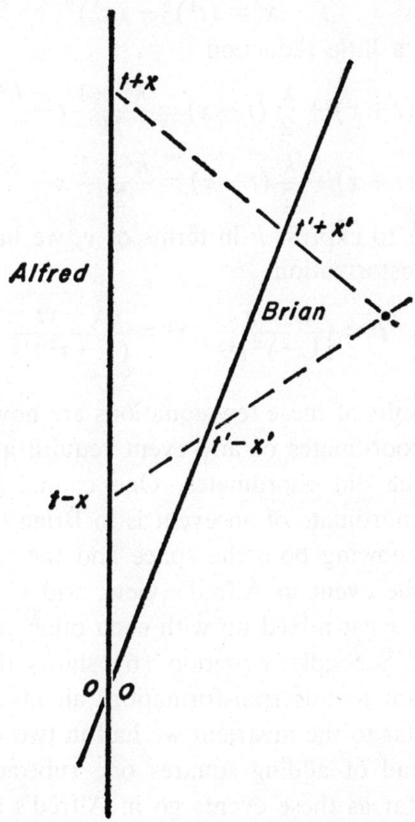

FIG. 23. *The Lorentz Transformation*

ceives it back at time $t+x$ so that he assigns coordinates t, x to the event. Similarly Brian emits a pulse at t^1-x^1 and gets it back at t^1+x^1. But in fact Brian emits his pulse as Alfred's pulse passes him and receives it as the returning pulse to Alfred passes him. Hence

$$\begin{aligned} t^1 - x^1 &= k(t - x) \\ t + x &= k(t^1 + x^1) \end{aligned} \tag{5}$$

Clearly

$$t^2 - x^2 = (t^1)^2 - (x^1)^2 \tag{6}$$

and, with a little reduction

$$t^1 = \frac{1}{2k}(t + x) + \frac{k}{2}(t - x) = \frac{k^2 + 1}{2k}t - \frac{k^2 - 1}{2k}x \tag{7}$$

$$x^1 = \frac{1}{2k}(t + x) - \frac{k}{2}(t - x) = \frac{k^2 + 1}{2k}x - \frac{k^2 - 1}{2k}t$$

Using (2) to express k in terms of v, we have the Lorentz Transformation

$$t^1 = \frac{t - vx}{(1 - v^2)^{1/2}}, \quad x^1 = \frac{x - vt}{(1 - v^2)^{1/2}} \tag{8}$$

The results of these few equations are now, first, that the new coordinates of any event require a knowledge of both the old coordinates. One cannot know what the time coordinate of an event is in Brian's view without first knowing both the space and the time coordinates of the event in Alfred's view, and vice versa, so that t and x get mixed up with each other in the transformation. Secondly, equation (6) shows that there is an invariant to this transformation, an invariant strikingly similar to the invariant we had in two dimensions, only instead of adding squares one subtracts squares. Thus, as far as these events go in Alfred's and Brian's plots, all the statements we made about why one called a plane two-dimensional apply now. Two numbers, t and x, are needed to specify the position of an event; they get jumbled up as we transform from Alfred's view to Brian's view, and there is an invariant connecting them. Therefore, one calls this space of Alfred and

Brian a two-dimensional space, in which time is one of the dimensions.

To examine the remaining space coordinates y, z and y^1, z^1 consider a light ray emanating from the event described by Alfred as t_0, x_0 and by Brian as $t_0{}^1$, $x_0{}^1$. Since light travels with unit speed, the journey of the flash is described by Alfred as

$$(t-t_0)^2-[(x-x_0)^2+(y^1)^2+(z^1)^2]=0 \qquad (9)$$

the square bracket being the square of the distance from the source. Similarly, in Brian's view

$$(t^1-t_0{}^1)-[(x^1-x_0{}^1)^2+y^2+z^2]=0 \qquad (10)$$

But

$$\begin{aligned}(t-t_0)^2&-(x-x_0)^2\\ &=[(t-x)-(t_0-x_0)]\,[(t+x)-(t_0+x_0)]\\ =\frac{1}{k}&[(t^1-x^1)-(t_0{}^1-x_0{}^1)] \qquad (11)\\ k\ &[(t^1+x^1)-(t_0{}^1+x_0{}^1)]=(t^1-t_0{}^1)^2-(x^1-x_0{}^1)^2\end{aligned}$$

Hence by (9) and (10)

$$y^2+z^2=(y^1)^2+(z^1)^2 \qquad (12)$$

and since any event is on some such flash, (12) is generally true. Since the y and z directions are symmetrically placed about the direction of motion we have the final two Lorentz formulae

$$y=y^1,\ z=z^1 \qquad (13)$$

Four Dimensions

The combinations of equation (6) and equation (13) then lead us to think of a space in which the three space coordinates x, y, z, can be jumbled up by a rotation; and the fourth coordinate t can be jumbled up with the others through a velocity by going from Alfred to Brian. If we consider the two processes, the rotation and the velocity, together, we arrive at a space of four dimensions, t, x, y, z, in the sense that one needs four coordinates to specify an event, namely, when and where it takes place. When one changes coordinates through a rotation, or through having a velocity relative to one's previous system, then these coordinates get jumbled up; and finally, there is an invariant, namely $t^2 - x^2 - y^2 - z^2$.

People have occasionally been baffled and frightened by this use of "four dimensions" and have thought that in some mysterious way physicists or mathematicians can imagine four dimensions. Nothing could be farther from the truth. All one means by four dimensions is that these four quantities, the time and the space coordinates, satisfy the conditions just stated and, therefore, one can treat them as dimensions in very much the same way one treats the ordinary space dimensions by themselves. Of course, what disturbed people at first was that, whereas previously they had thought of time as an invariant, it now turned out not to be so; but the only reason for not thinking of time as something that gets jumbled up with other coordinates was that no such high velocities had been considered. When the velocity is very small then the Lorentz transformation tells us that the time is unchanged. But time does not remain unchanged for large velocities.

Application of the Lorentz Transformation

A number of consequences of the Lorentz transformation can now be given—indeed, the whole subject of applied theoretical Relativity is simply the application of the Lorentz transformation.

First, the *Relativity of Simultaneity*. One sees straight away from the Lorentz transformation that, if Alfred regards two widely separated events as simultaneous, then Brian will not do so, for if these events are widely separated, then although Alfred will have assigned to them the same coordinate t, he will have given them different coordinates x, and by equation (8) Brian will then give them different time coordinates t. What one inertial observer regards as simultaneous happenings at spatially separated points, another will not regard as simultaneous. This is again an illustration of the private nature of time and a different derivation of this by the k calculus will be given in one of the later chapters.

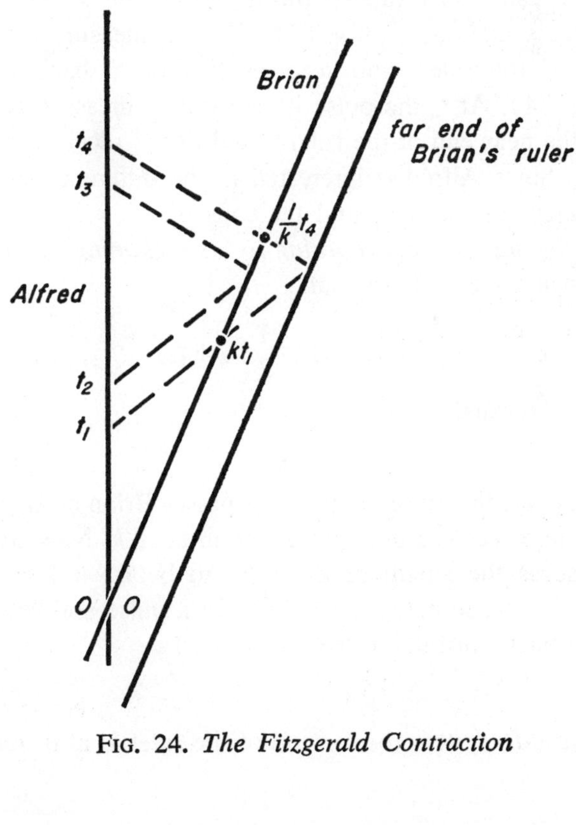

FIG. 24. *The Fitzgerald Contraction*

The Fitzgerald contraction is the name given to the fact that in Alfred's coordinates the length of a ruler held by Brian along Brian's direction of motion is shorter than in Brian's coordinates. This is immediately clear from the Lorentz transformation. Let the two ends of the ruler have coordinates, in Brian's system, $x^1 = 0$ and $x^1 = L$. By (8) this implies that Alfred's coordinates for them are

$$x = vt, \ x = vt + L(1 - v^2)^{1/2} \text{ respectively.}$$

Considering the two ends at the *same value of Alfred's time t,* Alfred thus ascribes a length of only $L(1 - v^2)^{1/2}$ to the ruler.

We can also establish this result in the k calculus. Alfred sends out a pulse at time t_1, to measure the far end of the ruler, and receives this pulse back at t_4 (Fig. 24). At t_2 the pulse is sent out to measure Brian (at the near end of the ruler), and this is received back at t_3. Since Alfred is interested in the difference in the distances of the near and far ends of the ruler *at one and the same time, according to his reckoning,* he must arrange his signals so that

$$\frac{1}{2}(t_1 + t_4) = \frac{1}{2}(t_2 + t_3) \tag{14}$$

Also, as usual

$$t_3 = k^2 t_2 \tag{15}$$

Moreover, the pulse emitted at t passes Brian at kt_1 and the pulse received at t_4 passes Brian at t_4/k. Now Brian measures the length of the ruler to be L, and so the time difference between sending out a pulse and receiving it back must be $2L$, so

$$t_4/k - kt_1 = 2L \tag{16}$$

In Alfred's system the far end of the ruler is at distance

$\frac{1}{2}(t_4 - t_1)$, the near end at $\frac{1}{2}(t_3 - t_2)$ and so Alfred measures the length of the ruler to be

$$\frac{1}{2}[(t_4 - t_1) - (t_3 - t_2)] \tag{17}$$

With the aid of (14), (15), (16), expression (17) becomes

$$\frac{2k}{k^2+1}L \tag{18}$$

which by (2) equals

$$L(1-v^2)^{1/2} \tag{19}$$

It will have been noted that the crux of the argument lies in Alfred's comparing the distances of the far end and the near end of the ruler at the *same time* in his reckoning. What Alfred *sees* is something quite different. If he took a snapshot he would see the far end of an earlier time than the near end, owing to the travel time of light (Fig. 25). If the two responses are to arrive at the same time t_3, the radar signals required to produce them will have to be sent out at different time t_1, t_2 and so reach Brian at kt_1, kt_2 respectively. Then $k(t_2 - t_1) = 2L$ (Brian's measurement), and so Alfred sees the length of the ruler to be

$$\frac{1}{2}(t_3 - t_1) - \frac{1}{2}(t_3 - t_2)$$

$$= \frac{1}{2}(t_2 - t_1) = \frac{1}{k}L\left(\frac{1-v}{1+v}\right)^{1/2} \tag{20}$$

a result quite different from (19).

To see the significance of (20), imagine Brian's ruler to hold lights at each end and to slide along a ruler at rest relative to Alfred. Then the two marks of Alfred's ruler that he *sees* illuminated at one and the same time will differ by (20). It is only when he makes allowance

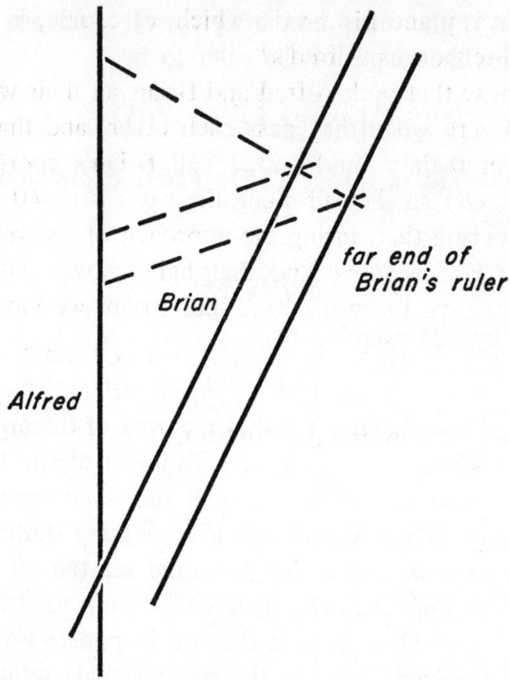

FIG. 25. *Looking at Brian's Ruler*

for the extra travel time of light from the far end of the ruler that he obtains the more sophisticated result (19).

THE ABERRATION OF LIGHT

Again, we think about Alfred and Brian with their x-axes aligned as before. A ray of light approaches Alfred and meets him at the very moment when Brian passes him. Alfred measures the direction from which the light came and finds a certain angle θ between that direction and his x-axis. Brian receives the same light at the same moment because he is just passing through Alfred's position; what angle θ' does he find between the light ray and his x-axis, which, of course, is in the same direction as Alfred's?

Suppose that both Alfred and Brian set their watches to read zero when they pass each other, and that they so orient their y- and z-axes (all Brian's coordinates are primed) so that all along the ray $z=z'=0$. Then, remembering that during the approach of the ray both t and t' are negative and that light travels with unit speed relative to both Alfred and Brian, we have

$$x=(-t)\cos\theta, \qquad x'=(-t')\cos\theta' \qquad (21)$$

$$y=(-t)\sin\theta, \qquad y'=(-t')\sin\theta' \qquad (22)$$

Substituting from the Lorentz formulae (8) into (21) we find by division

$$\cos\theta'=\frac{v+\cos\theta}{1+v\cos\theta}$$

Equivalent formulae could be derived by substituting (8) and (13) into (22).

We first notice that if $\theta=0°$, $\cos\theta=1$, $\cos\theta'=1$, $\theta'=0°$, while if $\theta=180°$, $\cos\theta=-1$, $\cos\theta'=-1$, $\theta'=180°$. But for all intermediate values there is substantial distortion. As an example, the following table gives the relation between θ and θ' for $v=0.8$, which corresponds to $k=3$, a case already considered:

$\theta \longrightarrow$	0°	30°	60°	90°	120°	150°	180°
$\theta' \longrightarrow$	0°	10°12′	21°47′	36°52′	60°	102°25′	180°

Note in particular that the entire region between $\theta=0°$ and $\theta=120°$, which for Alfred is three-fourths of the whole sky, is in Brian's view compressed to the region between $\theta'=0°$ and $\theta'=60°$, a mere one-fourth of his sky.

The upshot of all this is, then, that in the direction in which Brian is going he sees the world around him much more compressed than in Alfred's picture. If Brian is traveling very fast relative to Alfred, then what to Brian is a small section of the heavens surrounding the direction in which he is moving relative to Alfred, is most of the sky in Alfred's view; and what to Brian is the rest of the sky is to Alfred a small patch in the direction opposite to that in which Brian is moving relative to him. This is a very considerable distortion of the picture. This so-called aberration of light is of the greatest historical importance. If Brian, instead of being an inertial observer, were changing his speed, he would therefore see the heavens wobble around him—he would see the directions to stars make different angles between each other at different times. The Earth is such a non-inertial observer in its motion round the Sun; at one time it is proceeding in some direction in its orbit at 30 kilometers per second—one part in 10,000 of the speed of light; six months later it is moving in the opposite direction, and so there will be a discrepancy between the angles that the lines to different stars make at different times of the year. This discrepancy was, in fact, discovered by James Bradley in 1725, and he immediately called it the Aberration of Light.

The early explanation of the Aberration was an idea that if light were a bullet moving through the telescope and always coming from the same direction, it would make differently aligned holes in the back and the front according to the direction in which the telescope was moving. In fact, this gives quite a good answer, not very different from the relativistic one, but it is a basically wrong explanation. According to the old explanation, if one filled the whole telescope with water, then the speed of light in the telescope would be altered; the aberration should now be larger because the light would take longer from one end of the telescope to the other, and so the telescope would cover a larger distance while the light traveled down the tube. But this is not, in fact, so, although we know this only from rather indirect experiments. The relativistic answer that we have now derived from the Lorentz formulae fits the observations very well and, furthermore, it predicts that there should be no variation if the telescope were filled with water because aberration actually describes the direction from which the light comes.

The historic importance of Bradley's discovery was that it gave, for the first time, a direct proof of the Copernican idea that the Earth was moving round the Sun. What was observed, of course, was not the velocity of the Earth, which is unobservable, but the fact that it *changes* its velocity, that at different times of the year the aberration displaces stars differently. This evidence established the Copernican system beyond doubt.

CHAPTER XI

FASTER THAN LIGHT?

We have seen that, however often velocities less than the velocity of light are added to each other, the velocity of light is never reached, let alone exceeded. It is, therefore, a barrier. The things that move slower than light compose a whole class of objects–indeed, all those we are familiar with. Having found this barrier, we may speculate what might be on the other side of the barrier and what it would look like. What would the properties of particles moving faster than light be, if as familiar a word as particle can be used for something so utterly strange as the entity that would emerge?

Cause and Effect

Let us again think of Alfred and Brian. They are together at twelve noon on both their clocks and are separating at a constant speed, so that any interval of time registered by Alfred is seen as three-halves of that interval by Brian. We have discussed this case in several previous chapters and now make a further calculation.

Suppose that a hypothetical entity faster than light passes Alfred at 12:40 P.M. by Alfred's watch, moving in the direction of Brian (Fig. 26). The light that Alfred emits at 12:40 P.M. is received by Brian at 1:00 P.M. by Brian's watch, because of the ratio of three to two that specifies Brian's speed. Since this entity is supposed

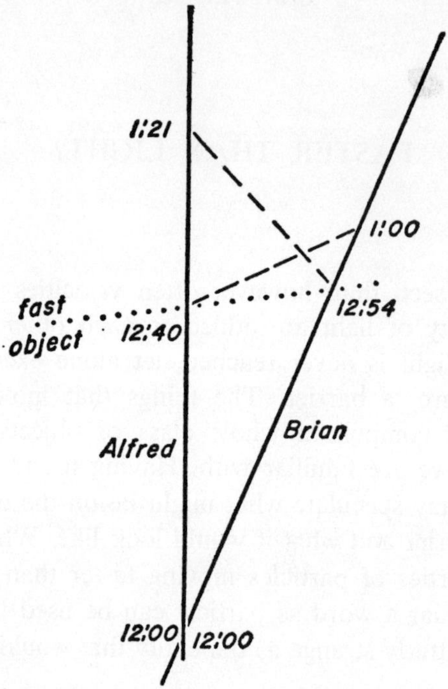

FIG. 26. *Alfred and Brian look at an object moving faster than light.*

to travel faster than light, it must get to Brian before Brian's watch indicates 1:00 P.M., to have a definite number at, say, 12:54 P.M. Alfred sees Brian's watch indicating 12:54 P.M., three-halves times 54 minutes after noon by his watch, that is, at 1:21 P.M. Thus, according to Alfred, this entity meets him at 12:40 P.M., and he sees it meeting Brian at 1:21 P.M.–later, that is. Brian, on the other hand, sees that entity meeting him at 12:54 P.M. by his watch and sees its meeting with Alfred at 1:00 P.M., that is, later. Thus, in Alfred's view, the entity first passes him and then Brian; in

Brian's view, it first passes him and later Alfred. Imagine any changes going on in this entity. If it is aging between its encounter with Alfred and its encounter with Brian, then Alfred will see it aging in the course of time like a normal object; Brian, on the other hand, will observe it getting younger. Which way then should this entity live? Which way does time run for it? Clearly, it would be something very peculiar for which the sense of time would appear to differ according to who looks at it. It would certainly upset all our notions of causality, of some events being causes for others. If something happened on this entity at its passage with Alfred and if this something then caused something else that occurred at its passage with Brian, Alfred would see cause precede effect, as we normally do; but Brian would see effect precede cause. Thus, in Brian's view this object would be in Looking-Glass Land, where the cake is first handed out and then cut and where, it will be remembered, punishment necessarily precedes the crime.

Of course, we cannot exclude that such very awkward things might occur. The physicist must always approach the world with an open mind. But he is allowed to heave a sigh of relief that no such entity traveling faster than light has ever been discovered. He would have to do a great deal of thinking to adjust himself to such a situation, though we might charitably have enough confidence in the flexibility of his mind to suppose that he could cope even with this. Fortunately, as we have been saying, no such curious entity has ever been discovered, and so we can stick to the ideas of causality, of causes preceding effects.

Enough has been said, however, to make it clear that any entity of this kind would really be so vastly different from anything we know that it is very fortunate that the velocity of light is such a perfect barrier separating the known type of object from those that are essentially unknown and, we may hope, will not be met for a long time.

Simultaneity of Spatially Separated Events

A related problem is that of whether two distant events are simultaneous. We are used to regarding this question as absolute. We tend to think that if two occurrences are simultaneous, however far they may be separated spatially, then they are simultaneous for whoever looks at them. But this would not be so when fast-moving observers were concerned. Let us again consider our friends Alfred and Brian and, to make things definite, say that Brian is moving west relative to Alfred, again with such a velocity that the ratio of the intervals of transmission and reception is three-halves. Imagine then two occurrences to which Alfred ascribes the distance of one light hour, both occurring at twelve noon in Alfred's reckoning, one to the east of him and one to the west of him (Fig. 27). When we analyze this statement, it means that if Alfred sends out radar pulses at 11:00 A.M. in both directions and if they are reflected by these simultaneous events one light-hour away, the pulses will both arrive back at the same time, at 1:00 P.M. Hence he would deduce their distances to be one light-hour from him and the events to have been at twelve noon in his reckoning–one to the east and one to the west.

Now let us look at these occurrences from Brian's point of view (Fig. 28). At twelve noon when in Alfred's reckoning all this happened, Brian was just passing him. First, consider the radar pulse Alfred sent out at 11:00 A.M. to the eastern event. This signal will pass Brian, since Brian before noon is to the east of Alfred

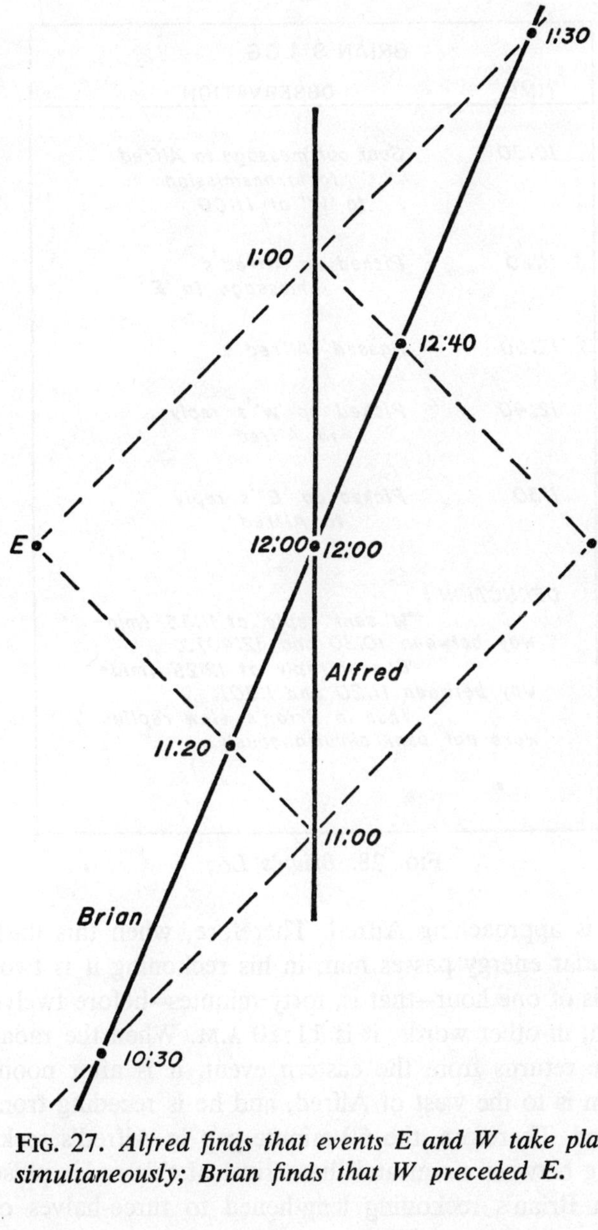

FIG. 27. *Alfred finds that events E and W take place simultaneously; Brian finds that W preceded E.*

BRIAN'S LOG

TIME	OBSERVATION
10:30	Sent out message to Alfred for transmission to 'W' at 11:00
11:20	Picked up Alfred's message to 'E'
12:00	Passed Alfred
12:40	Picked up 'W''s reply to Alfred
1:30	Picked up 'E''s reply to Alfred

DEDUCTION:

'W' sent reply at 11:35 (mid-way between 10:30 and 12:40).

'E' sent reply at 12:25 (mid-way between 11:20 and 1:30).

Thus in Brian's view replies were not sent simultaneously.

FIG. 28. *Brian's Log*

and is approaching Alfred. Therefore, when this flash of radar energy passes him, in his reckoning it is two-thirds of one hour—that is, forty minutes—before twelve noon; in other words, it is 11:20 A.M. When the radar pulse returns from the eastern event, it is after noon, Brian is to the west of Alfred, and he is receding from Alfred. Therefore, the 60-minute gap in Alfred's reckoning between noon and the arrival of this radar pulse is in Brian's reckoning lengthened to three-halves of that time, that is, 90 minutes. Thus on its return journey the radar pulse from the eastern event reaches Brian at 1:30 P.M. In other words, if Brian had emitted a radar pulse at 11:20 A.M. it would have illuminated this event, been reflected, and returned to Brian at 1:30 P.M. Hence Brian ascribes to the eastern event a time halfway between emission and reception, that is, 12:25 P.M., and its distance would be described as half the difference in these times, that is, one light-hour and five light-minutes.

Now consider the western event. When would Brian have had to send out a radar pulse to illuminate this event? Before noon Brian was to the east of Alfred and his radar pulse would have had to pass Alfred at 11:00 A.M. in Alfred's reckoning, so as to travel in company with Alfred's own pulse. When the three-halves rule is applied again, it follows that Brian would have had to emit this pulse 90 minutes before noon, that is, at 10:30 A.M. When the pulse was returned then Brian was to the west of Alfred, and, therefore, the pulse would reach him before it reaches Alfred.

The 60-minute gap in Alfred's reckoning would be only two-thirds as long for Brian, and so the radar pulse would arrive back at Brian at 12:40 P.M. Thus Brian would assign to the western event a timing of 11:35 A.M. and a distance of one light-hour and five light-minutes. Hence the two events, simultaneous in Alfred's reckoning, would not be simultaneous in Brian's reckoning. For spatially separated events, simultaneity is a relative and not an absolute property.

We can now analyze the time concept a little more closely. In ordinary life we are very well used to a particular property of time, namely, that it is ordered. Whenever two things happen, we can say either that one happened before the other, or the other way round, or that both happened at the same time. But what we have now established has rather upset this situation. For a distant event and one close by, different observers may evidently have different ideas of time; they may judge the events to be simultaneous, or one way round in time, or the other way round in time. This is disturbing. And so we next ask whether the whole idea of "before" and "after" is meaningless. Fortunately, the answer is no.

Suppose I do one thing today and another tomorrow. Then imagine other observers looking at these acts. Since there is never any overtaking of light by light, the rays emitted by my earlier action will be received by every observer earlier than my rays emitted at the time of the later action. So, as far as I am concerned, all observers will judge the two actions to be in the same order in which I performed them myself. For some events, evidently, the "before" and "after" idea holds universally. It is absolute and not relative, whereas for others—as we saw in the last example—this is not the case. Some pairs of events are absolutely ordered in time in the sense that every observer will agree which of the two happened first. But other pairs of events are such that different observers will have different views as to which happened first and which second. Where does the boundary between these two classes lie? Evidently the question of the spatial separation of the events is involved. The events that we considered in the example with which we established the relativity of simultaneity were widely separated. The events which everybody agreed happened in the same order both happened to me.

PAST AND FUTURE: ABSOLUTE AND RELATIVE

Let us again consider two observers, Alfred and Edgar. We say nothing at all about their relative motion and relative position except that at the material times of this example they are not together. Something happens to Alfred and, at that instant, he sends out a flash of light. This flash of light is received by Edgar at, say, noon on Edgar's watch. An hour later, Edgar sits down to write a letter. We have three events altogether in this example: Alfred's sending out a flash of light; next, Edgar's reception of the flash of light; and, third, Edgar's sitting down to write his letter. Since Edgar did both the receiving and the writing, it follows from what we said earlier that every observer will agree that Edgar did not start to write the letter until after he had received the flash of light from Alfred. Furthermore, every other observer could have followed the flash of light as it traveled from Alfred to Edgar. Whatever their state of motion, all observers would at least agree that Alfred emitted the flash before Edgar received it. Thus, the three events are strictly ordered. There is no question but that all observers, however they might be moving, would agree that Edgar started writing his letter later than the moment when Alfred emitted the flash of light. Therefore, we may regard the writing of the letter as absolutely later than the emission of the flash.

We can go a bit further. Only light, traveling at the speed of light, could get from Alfred's emission of the flash to Edgar's reception of the flash, but this restriction does not hold for Edgar's next event, his sitting down to write a letter. Whatever Edgar's speed may be relative to Alfred, it will be less than that of light, as was shown in Chapter IX. Therefore, while light from Alfred's event would get to Edgar an hour before his sitting down to write, in theory Alfred, when he emitted the signal, might also be able to fire a bullet (a hypothetical bullet, of course) that would not reach Edgar until the exact instant of writing. The bullet, obviously, would have to travel at a very high speed, indeed, but it still would be a theoretically *possible* speed, since it would be less than the speed of light. So we can say that any event involving Edgar after the arrival of the flash from Alfred is absolutely later than the emission of the flash from Alfred and could have been reached by a particle traveling from Alfred to him at a speed less than that of light.

Now think of Edgar's past. If somewhere in his past history Edgar had emitted a flash of light, it would have reached Alfred at some time. Suppose the flash reached Alfred at the instant Alfred emitted the flash we have been talking about. Then, since all inertial observers can see that flash of light traveling from Edgar to Alfred, they would all have agreed that Edgar emitted this flash before Alfred received it. If Edgar had done any particular thing (say, eaten his lunch) before emitting that flash to Alfred, then all inertial observers however traveling would have agreed that he had had lunch before he emitted the flash–that is, before Alfred received Edgar's flash and thus before Alfred emitted his own flash. Therefore, we can say that the part of Edgar's history before he emitted the flash occurred before Alfred emitted his; and all events that took place in this period of time may be called absolutely earlier than Alfred's emission of his flash.

Thus we have found two moments in Edgar's life that are singled out in relation to Alfred's emission of his flash, which for simplicity we shall call event X. There is the instant P at which Edgar would have to press the button of his light to illuminate X, and the instant F, at which Edgar receives the light emitted at X. All Edgar's life before P, as we have seen, will be regarded as *earlier* than X by every inertial observer. All Edgar's life after F will be regarded as *later* than X by every inertial observer. In Edgar's experience P evidently precedes F, and so one wonders whether the stretch of his life between P and F should be regarded as earlier or later than X. There is in fact no absolute and universal way of deciding. If we consider an event in Edgar's life later than P, but earlier than F, then some inertial observers will regard N as having happened *before* X; others will view it as having occurred after X, and yet others will regard X and N as having taken place simultaneously. Therefore, in relation to X the stretch between P (the end of the "absolute past") and F (the beginning of the "absolute future") is called the relative past and future.

It is not difficult to fit an inertial observer into our space-time diagram (one dimension of space only!) who will find X and the event N to be simultaneous. First we draw, from the view of some inertial observer, the event X and a line representing Edgar (Fig. 29). Next we draw the two dashed lines (both at 45° to the vertical) that represent the light rays illuminating X and emanating from X. They intersect Edgar's line in P and F respectively. Then choose some event N on Edgar's line between P and F. Now we can construct the line of the inertial observer passing through X who regards N as having occurred simultaneously with X. First we draw the light rays AN, NB illuminating N and emanating from N respectively. Then with X at the center we draw a circle passing through N. This circle intersects AN, NB in A′, B′ respectively. Note that, because of the right angle at N, the three points A′, X, B′ are in line,

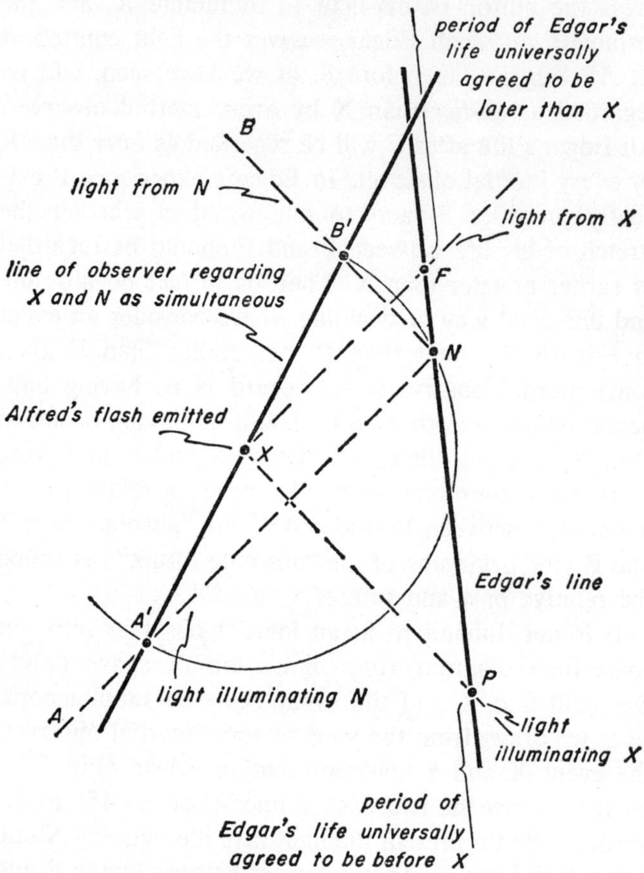

FIG. 29. *Alfred's line is not shown since only the event X on it is relevant here.*

and that this line is necessarily more nearly vertical than one representing a light ray. Therefore A′B′ through X represents an inertial observer. If this observer wants to illuminate N, he must press the button for his flash at A′, and he will receive the answering flash at B′. He thus regards N as having happened halfway in time between

A′ and B′, that is, at X. Hence he finds that N and X were simultaneous.

Clearly then, we can find other observers who will find event N to have taken place before event X (Alfred's emission of the flash), and yet others will have found it to be the other way round. Thus, we can say that in relation to X, Edgar's life falls into three parts. One, before the moment P when Edgar would have to emit a flash of light to illuminate X, is the absolute past. Everybody will agree that any such event involving Edgar in this absolute past occurred before X. Part two falls between P, the end of Edgar's absolute past, and F, when Edgar receives the flash from Alfred. In relation to time this period cannot be ordered absolutely with Alfred's flash. For any event in this patch of Edgar's life, some people will say that it occurred before Alfred's emission of the signal, some that it occurred later, and some that it occurred simultaneously. This whole stretch will, therefore, be referred to as Edgar's relative past and future, because relative to some observers it will be before Alfred's event X, and relative to others it will happen after Alfred's event. The third period of Edgar's history is after F, the reception of the light ray from Alfred, because all observers will agree that whatever Edgar did during this period would have happened after Alfred's flash was emitted. This stretch may be referred to as Edgar's absolute future. We now consider other observers to the right of X in addition to Edgar. Just as in Edgar's case, the world line of each such observer will contain an instant P marking, in relation to Alfred's event X, the end of his absolute past, and an instant F marking the beginning of his absolute future. For every such observer P occurs when the flash of light illuminating X passes him, while F occurs when the light emitted at X reaches him. Thus these two

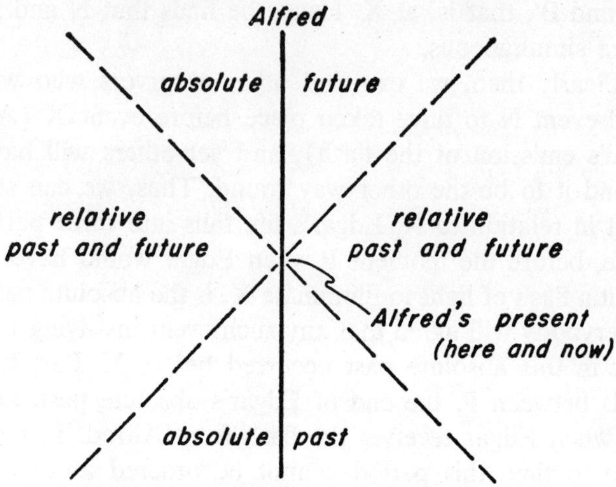

FIG. 30. *The Light Cone in Two Dimensions*

flashes of light are the boundaries of the three regions (absolute past, relative past and future, absolute future). What we have considered in relation to observers to the right of Alfred's event X naturally holds also to the left of X. Therefore (Fig. 30) the light rays that Alfred emits, together with the light rays that illuminate Alfred at the moment of emission, divide the whole of space and time into three separate parts: the absolute past, the absolute future, and, between them, the relative past and future.

THE LIGHT CONE

In Fig. 30, as in all our previous diagrams, we have made use of only one dimension of space, using the second dimension of the paper to represent time. Let us now try to use two dimensions of space. We shall have to make a perspective drawing (Fig. 31) to represent three dimensions (two space and one time) on the

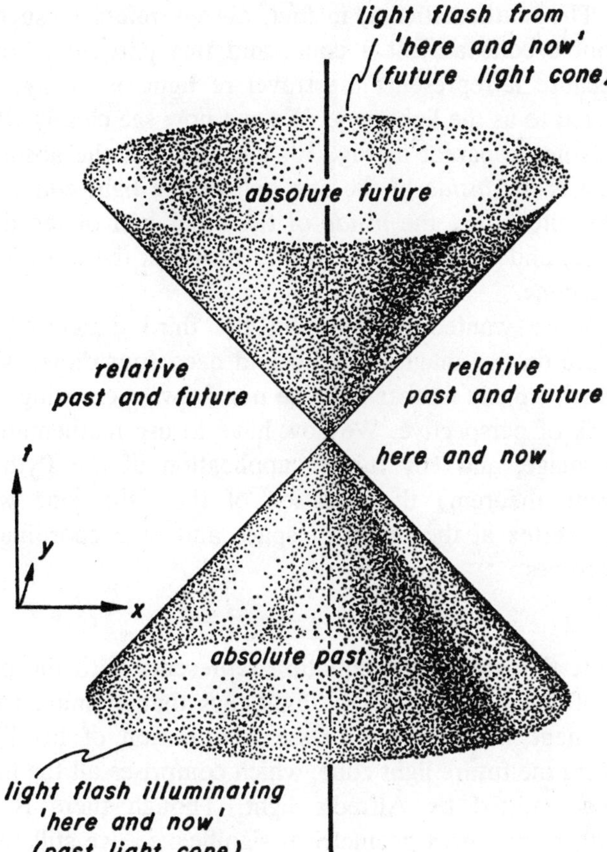

FIG. 31. *The Light Cone.*

paper. The light rays emitted by Alfred's event now form a cone with its vertex at Alfred. Similarly, all the light rays that would arrive at Alfred just at the moment of Alfred's event would form a second cone, again with Alfred at the vertex—the generators of the two cones are identical and they all meet at the common vertex.

The mathematician, in fact, always refers to such a double cone as just a cone, and this particular cone, because it represents the travel of light, is always referred to as the light cone. We can now see clearly what was not apparent in Fig. 30; namely, that the absolute past is the *inside* of the one half of the light cone, the absolute future the *inside* of the other half of the light cone, and the relative past and future is the *outside* of the cone.

If we wanted to represent the third dimension of space on our diagram, we would need four dimensions which we can neither imagine nor draw, even using any trick of perspective. We now have to use mathematical language, and (by further application of the Pythagorean theorem) the equation of the light cone with the vertex at the origin of space and time coordinates becomes

$$t^2 - x^2 - y^2 - z^2 = 0$$

Note that this single equation represents both the past light cone—that is, all the light rays that illuminate the moment when Alfred presses the button of his light—and the future light cone, which comprises all the light rays emitted by Alfred's light. Though there is no longer any direct geometrical significance, we still refer to the surface defined by this equation as the light cone, with its two halves—past and future—and we still say, with the same justification as before, that the inside of the past light cone is the absolute past, the inside of the future light cone is the absolute future, whereas the outside is the relative past and future and cannot be ordered in any absolute sense relative to the origin.

Return now to the notion of cause and effect. If we accept the idea that there are things that may be caused and that the cause must precede the effect, then it follows that every cause of an effect must lie within or on the past light cone of the effect. Similarly, we must then assume that any effect caused by something happening here and now must lie on or within the future light cone. If this were not so, if there were cause/effect links running outside the light cone from here and now, then the order could be reversed by viewing the situation from a suitably moving observer. We, therefore, arrive at the result that, within the context of our ideas of cause and effect, no influence can travel faster than light, just as we have seen before, as a matter of observation, that there is no body traveling faster than light.

It would be a mistake, however, to say that nothing happens faster than light. If there is no cause and effect link and no material body concerned, then we may well have something traveling faster than light and we can easily show this. Suppose we have two long straight rulers, one on top of the other, inclined to each other at a very slight angle; then, if we move one of the rulers at right angles to itself, the point of apparent intersection will move along the rulers. For a given speed of the moving ruler, however slow, the point of intersection will move as fast as we like if the angle between the rulers is sufficiently small. Thus we can always find an angle so small that the point of intersection goes at a speed exceeding the speed of light. There is nothing in Relativity against this. For what travels there faster than light is neither a material particle nor an influence, but a geometrically defined point which cannot move anything and cannot do anything.

We can also imagine, as another example of speeds greater than light, a searchlight of enormous power mounted on the earth. We now turn this searchlight; the beam will sweep across the sky, far far away where it meets the planets or perhaps even the stars. The end of the beam will travel at speeds vastly greater than that of light because of the huge length of the arm of the lever which we are turning gently on the earth when we change the direction of the searchlight. But, once again, this is no material point; this is no influence that is hopping from planet to planet much faster than light. This is merely a succession of events that just happen to be connected because we happen not to switch off the searchlight at moments during the turning. But what we have established is that there can be no link between cause and effect that travels faster than light, just as we have discovered that there is no object moving faster than light.

CHAPTER XII

ACCELERATION

In previous chapters we have considered three observers, Alfred, Brian, and Charles, and we have had Brian and Charles moving with the same velocity relative to Alfred but in opposite directions. The meeting between Charles and Alfred occurred later than the meeting between Brian and Alfred; but the timing by Brian and Charles of the period between these two meetings gave a lower value than the timing by Alfred. Very brief reference was then made to what would happen if Brian, on passing Charles, threw his young son across and Charles caught him. However, this question demands far closer and more detailed analysis.

Up to now, with the brief exception of the reference to this boy, we have always been talking about *inertial* observers only—that is, observers moving with constant velocity. In the Principle of Relativity reference was made only to such inertial systems, and all the statements concerned such systems. The Principle of Relativity asserts the impossibility of distinguishing one inertial observer from any other by internal experiment. Now, however, we wish to discuss systems in a state of acceleration—that is, systems that are not inertial.

The Principle of Relativity can be taken to say that velocity does not matter, and this statement is complete and all-embracing. On the other hand, acceleration does matter and it is not so easy to specify just how it matters. To say that something does not matter is a complete description; to say that something does matter requires a great deal of further elucidation.

That acceleration matters is indeed obvious, as it is an everyday experience. Sitting in a car that is accelerated, say, by rapidly increasing its speed, we feel ourselves pressing hard against the backs. We feel this because the car subjects us to an acceleration and, therefore, must exert a force on us. Thus the absolute significance of acceleration cannot be denied. How significant it is depends on its magnitude, and just what effect its magnitude has depends on the construction of the object that is subjected to such an acceleration.

Acceleration and Clocks

We have been much concerned with time-keeping, and so we consider the effect of acceleration on clocks and watches. An ordinary wristwatch is not significantly affected by gesticulating while one talks. It is swung about on one's arm, but the accelerations are too small to affect the mechanism—at least, it is not affected seriously. The same watch dropped on a concrete floor suffers a considerably higher acceleration on hitting the floor; this acceleration (you will recall that any change of speed or direction is an acceleration) will usually break the watch. There is accordingly a limiting acceleration for this kind of watch. If only smaller accelerations occur, as in gesticulating, there is no serious effect, but larger ones break the watch. A shockproof watch can take higher accelerations, such as being dropped from table-height onto a concrete floor; but if it were dropped, say, from the top of the Eiffel Tower, attached to a massive object so that air resistance did not matter, then, no doubt, it would disintegrate on hitting the ground. Thus for a shockproof watch there also exists a limiting acceleration, higher than in the previous case, but finite nevertheless.

We next consider even tougher clocks. Perhaps the most solid time-keeping mechanism that one can readily think of is a radioactive material such as, say, radium. Radium disintegrates spontaneously with a half-life of about 1620 years; that is to say, after 1620 years half the material will have disintegrated. After a further period of 1620 years half the remainder will have disintegrated, and so on. Radioactive materials with much shorter or much longer half-life can be chosen to suit any time-keeping purpose. This rate of disintegration is quite independent of what happens to the material, within wide limits. We can subject it to hammer blows, put it into sticks of dynamite, and explode it, heat it to millions of degrees, cool it as near absolute zero as we can, and its half-life will be unaffected. Thus it can be used as a time-keeper over a vast range of circumstances with great accuracy. We can subject it to very large accelerations indeed without affecting the time-keeping mechanism locked up in the nucleus. If, however, we wanted to subject it to really enormous accelerations, then we could do this only by bombarding it with very fast particles. They, and they alone, could be used to convey these very high accelerations; but if the incoming particles were sufficiently energetic they could disintegrate even the nucleus of radium. Thus, this clock, too, has a limiting acceleration although it is exceedingly high.

We need not think only of physical clocks. We can use biological ones just as well. We can use the generations of rabbits or the reproductive cycle of the sea urchin. Subjecting the rabbits to minor accelerations, as on a car ride, will not seriously interfere with them, but if we subject them to really high accelerations they are liable to die, and similarly for the sea urchins.

We ourselves, too, could be used for time-keeping. Our aging is a time-keeper, and so is the frequency with which our stomachs tell us that we are hungry. Because there are large individual differences we are perhaps not very good time-keepers, but we are time-keepers nevertheless. Again, there are limiting accelerations; if we are subjected to an acceleration of one or two *g* we are all right, though, conceivably, we might feel seasick. If, on the other hand, we were subjected for appreciable periods to accelerations of twenty *g,* we would die, and our time-keeping would cease. Note that as long as any of these time-keeping mechanisms move at constant speed (that is, as inertial observers) they all indicate the same time. For if this were not so it would contradict the Principle of Relativity, as we would have a means of distinguishing between inertial systems. According to the Principle of Relativity there is *no* means of distinguishing between them, and so this ratio of times indicated by our different clocks must be identical for all inertial observers.

The Twin "Paradox"

In the experiment of Alfred, Brian, and Charles, all three were inertial observers, but the boy who was thrown across from Brian to Charles was not. He suffered a period of acceleration; if this was as quick a transfer as the circumstances warranted, it would, no doubt, have led to the boy's immediate death; if, however, instead of his son, Brian had used a packet of a suitable radioactive substance, then its disintegrating properties would have served as a perfect time-keeper for Charles after the transfer in spite of the high acceleration involved. If, in that experiment, we changed all the times from hours to years, we could then afford to make the transfer from Brian to Charles more slowly (that is, with lower acceleration) and we could arrive at a situation where even a living being could survive the acceleration. In this form the example used to be called the twin paradox. Alfred and Brian were regarded as twins living together; then Brian began to move with an acceleration that he could survive, changed his velocity by a further acceleration to return with Charles, and then, by yet another period of acceleration, came to rest next to Alfred.

As we have seen, the Brian/Charles measurement of time is lower than the Alfred measurement of time. Thus Brian would finally again be living with Alfred, but would not have aged as much as Alfred. They would be twins of different ages. Of course, it was always ridiculous to call this a paradox; no paradox of any form is involved, for Brian has undergone several periods of acceleration in his life, whereas Alfred has been inertial all the time. It has sometimes puzzled people how, with relatively short periods of acceleration, Brian could have lost all this time. How can all this time have got sunk, as they said, in the relatively short periods of acceleration? But this argument is wrong; it is based on the hidden assumption that somehow Brian has "lost" time. Nothing of the sort has happened; Brian has measured *his* time and Alfred *his* time and there is no reason to believe that the two should be the same. There is no universal time, because time is a route-dependent quantity.

The situation is completely analogous to that of driving from one town to another. The shortest route is a straight line; if somebody travels a long route consisting of two straight lines joined by a short and sharp curve, then the second driver will have covered a larger mileage because there is a curve on his route; but the extra mileage will not lie *in* the curve, it will be *due* to it. There is only one shortest road, because there is only one straight line between two points; any other line must necessarily have at least one bend in it. The extra mileage then is due to this bend without necessarily in any way residing in the bend. Similarly, the shortness of the time between the first and the last meeting of our observers, as measured by Brian, is *due* to Brian's having undergone periods of acceleration, but in no way can we say that time has been standing still for him (or going backward) *during* these periods of acceleration.

There is only one way of getting from the first meeting to the last without acceleration—namely, the inertial mode of travel followed by Alfred. Any other way of getting from the first to the last meeting involves accelerations, and this means that the time taken by such an observer is less than the time recorded by the inertial one.

How Far Can We Travel in Space?

This raises a fascinating little question. How far can we travel in space, subject to our biological limitation? We want to leave out of consideration all the extremely serious technological limitations that restrict space travel even in our day of advanced technology. On the other hand, we want to confine ourselves to accelerations that we can bear, and to lengths of travel time, as measured by the traveler, that we can survive. Suppose we travel in a space ship that is always subject to an acceleration g. This is just the same as the gravitational field that the Earth produces around us. Hence life in this space ship would be very comfortable. We would attain very respectable velocities in the course of a few years, very close to the velocity of light, and thus we can usefully employ this mode of travel.

Suppose we start off from here with acceleration *g* for a certain period, say, 10 years of our lives. We then reverse the direction of our rockets and subject ourselves to the same acceleration but in the opposite direction for a period of 20 years by our reckoning. The changeover may be momentarily disagreeable, but we do know that this kind of thing will not do any permanent harm to us. Having attained a certain speed relative to our starting point in the first 10 years, we will need the next 10 years of opposite acceleration to reduce this motion to rest relative to the starting point again, and then a further 10 years to bring the rocket to the same speed in the opposite direction. Switching the direction of the acceleration again, we will find that the final 10 years will bring us back to rest on the Earth. Thus we will have aged 40 years in this journey, about as much as we conveniently can during our working lives.

Seen from the Earth, however, we have been moving with terrific velocity, so much so that for most of the time we have been traveling at almost the speed of light. In fact, as observers on the Earth see it, the farthest point reached in our travels turns out to be 24,000 light-years from the Earth. Of course, the people on the Earth have noted the passing of much more time than we in our travel at such high speed relative to the Earth. We come back to quite a different situation; to an Earth 48,004 years older than when we left it. Perhaps few of us would like to undergo such an experience, but, nevertheless, it gives one an idea of what we are biologically capable of. Thus we can in this way travel to places in space about 24,000 light-years away, about the distance to the nucleus of our own galaxy, though not nearly as far as any other galaxy.

If we are capable of taking 2*g* for forty years then we could travel to distant galaxies over 600 million light-years away, and would return correspondingly to an Earth over 1200 million years older. That we come back only to tell a much later generation is a serious matter, as is the fact that our most advanced rocket engineers could not dream of producing a rocket capable of maintaining such accelerations for the periods in question. This limitation, however, is a matter of technology, not of biology.

CHAPTER XIII

PUTTING ON MASS

We started our investigations with Newtonian dynamics, which led to the Newtonian Principle of Relativity. When optics, and with it the concept of light as a fundamental unique entity, was added, Einstein's Theory of Relativity followed. We then worked out various consequences of this theory, always using the technique of light signaling. In this chapter we return to dynamics to see how Newtonian dynamics, known to apply at low speeds, must be modified to fit in with Relativity when high speeds are considered.

THE STRETCHING OF TIME

First, we have to look more closely at the notion of time. We return once again to our friends Alfred and Brian, both inertial observers, both at the same place at 12 noon according to both their watches, but moving at such speed relative to each other that after this meeting there is a ratio of three to two of the interval of reception to the interval of emission. Therefore light rays sent out by Alfred 40 minutes by his watch after their meeting will reach Brian at 60 minutes by Brian's watch after their meeting; Brian instantaneously returns this light, and Alfred will receive it 90 minutes after their meeting.

Thus, if Alfred wishes to assign a time to the moment

when Brian returned the light, it would be halfway between 40 and 90 minutes after their meeting, which is 65 minutes. Though this moment is only 60 minutes after the meeting by Brian's watch, Alfred finds it to be 65 minutes after their meeting. The important point here is not that Alfred sees Brian's watch indicate 60 minutes after the meeting—that is, 1:00 P.M.—when it is 1:30 P.M. by Alfred's watch, but that Alfred's only way to allow for the travel time of light is to take the mean between the time of emission and the time of return of his pulse, which gives him 1:05 P.M. Even then his time does not agree with Brian's, and even when he has thus made allowance for the travel time of light, Brian's watch still seems to him to be going slow, to have advanced only by 60 minutes in 65 minutes of his own time. Naturally, if Brian were to move faster, this effect would be more pronounced. Thus if we take again, as we have done once before, the case when the ratio between the times of emission and times of reception is three to one, then light that reaches Brian at 1:00 P.M. by Brian's watch must be sent out when it is 12:20 P.M. on Alfred's watch. The returning light would arrive at Alfred at 3:00 P.M. by Alfred's watch. The mean between those two, 12:20 P.M. and 3:00 P.M., is 1:40 P.M., and this is the time assigned by Alfred to the moment of reflection, 1 P.M. by Brian's watch. Thus, in Alfred's reckoning, when he has allowed for the travel time of light, it again is true that Brian's watch is going slow, having covered only 60 minutes in 100 minutes (from 12 noon to 1:40 P.M.) of Alfred's time. This is the stretching of time or time dilatation, which in the second case (to be used throughout the rest of this chapter) is in the ratio one hundred to sixty (five to three).

Next, let us suppose that Brian has a ruler marked

in feet and held at right angles to the direction in which he sees Alfred, and that Alfred has a similar ruler held parallel to Brian's, and therefore also at right angles to the direction in which he sees Brian. Then a foot on Brian's ruler looks a foot to Alfred and vice versa, for if Alfred moves a foot along his ruler, then in the direction at right angles to his ruler there will be the foot mark on Brian's ruler. Thus there is no difficulty in translating distances along these two rulers into each other–they are simply the same.

Now let Brian have a particle which moves along Brian's ruler at a speed, in Brian's reckoning, of 60 miles an hour. What will this speed look like to Alfred? In a given time of Brian's a certain distance is covered on Brian's ruler and Alfred will entirely agree about this distance; there is no difficulty in translating it. But the time that in Brian's reckoning is only 60 minutes is 100 minutes in Alfred's reckoning.

It therefore appears to Alfred that this particle, instead of moving at 60 miles an hour, is moving at only 36 miles an hour, because the 60 minutes of Brian's time that the particle took to cover 60 miles appeared to Alfred to have been 100 minutes, and covering 60 miles in 100 minutes corresponds to a speed of 36 miles an hour. Thus what Brian observes as one velocity along this ruler will appear to Alfred to be only 60 percent of that velocity, lengths being unchanged but times appearing dilated.

Increasing Mass

The fundamental dynamical quantity, however, is not velocity but momentum. You will recall that this important concept of Newtonian dynamics is the product of velocity and mass, and satisfies a conservation law.

For our present purposes, we want to consider a simple measure of momentum. For example, we might measure the momentum of a bullet by the greatest thickness of armor plating it can penetrate. We may suppose that the process of penetration is wholly determined by the momentum of the bullet at right angles to the sheet of armor plating, and leave out complicating factors such as the shape and material of the bullet. If Brian fires a particular type of bullet at a particular speed, he can then determine the maximum thickness of armor plating that this bullet can penetrate. If, in Brian's view, the armor plating is at right angles to the path of the bullet, then his measure will be in the right direction to show the thickness of the armor plating. Since Alfred sees the armor plating edge on, he will agree with Brian's measurement of its thickness. Thus Alfred can deduce the momentum of Brian's bullet and gets the same value as Brian. On the other hand, Alfred's measurement of the velocity of Brian's bullet along the transverse measure yields only 60 percent (three-fifths) of Brian's value. In order to arrive, as required, at the same momentum as Brian, Alfred deduces for the mass of the bullet five-thirds of Brian's value. This enhancement of mass is clearly due to the time dilatation, which in turn is due to Brian's velocity relative to Alfred. Hence Brian's velocity has, for Alfred, increased the mass of Brian's bullets. This effect must enhance the mass of *all* Brian's objects, since they could all be employed in the manner of the bullets, or equally the bullets could have been used to calibrate Brian's scales.

This increase of mass (i.e., of inertia) is easily related to another quantity. In Alfred's view, Brian's motion relative to him means that Brian and all his accompanying objects have a substantial energy of motion (kinetic energy). Dividing this energy by the square of the velocity of light, one arrives essentially at the extra mass for each of Brian's objects. In our units, with velocity of light equal to one, and for moderate velocities where Newtonian dynamics applies, the extra mass is just equal to the Newtonian kinetic energy

$$\tfrac{1}{2}mv^2$$

It is thus reasonable to suppose that here, as elsewhere, Relativity gives the extension of Newtonian terms to high velocities so that the extra mass always just equals the kinetic energy *in our units*. To translate to general units, we observe that energy is given in terms of mass, length, and time, by the product of mass and the square of a velocity. Thus in units in which the velocity of light is not one, but say, c, the extra mass equals the kinetic energy divided by c^2. Thus our result can be interpreted, with equal justice, either as an increase of mass with velocity or as a mass of energy, in which we regard the extra mass of the bullet as being the inertia of its energy of motion.

Following the first interpretation we see thus that mass, instead of being a constant as in Newtonian theory, becomes something velocity-dependent in Relativity, and it is easily seen that the mass becomes arbitrarily large if the speed is increased to sufficiently close to the velocity of light. It is worth noting, though, that if the "proper speed" introduced in Chapter IX is used, the momentum is obtained by multiplying it by a velocity-independent mass.

When a particle is moving at a speed not much below the velocity of light, its mass is very much greater than its mass when it is at rest, the so-called rest mass. Putting more energy into the particle so that it hits things harder can then increase its speed only very little but does increase its energy and hitting power by making it more massive, and there is no limit to this process. The detailed and precise observations of this effect constitute perhaps the best test of Relativity.

Accelerating Protons

As an example, we may consider the protons moving in the huge joint European Particle Accelerator (CERN), at Meyrin, near Geneva. The energy communicated to these protons at the top of the performance of this, the world's biggest accelerator, is 28 BeV in the units nuclear physicists use.[1] At this energy the mass of the protons in the sense discussed earlier is around 30 times their rest mass. They move at a speed that is less than the velocity of light (186,000 miles per second) by only 100 miles per second. As a further illustration, consider what happens as the protons are being accelerated to this speed. When they have 95 percent of their final energy, their speed differs from their final speed by only one part in 18,000 of the velocity of light—i.e., about ten miles per second. This extra speed is, in fact, not very different from the maximum speed of the space probes sent up in recent times. The final increase in energy therefore adds relatively little to the speed of the protons. Its real purpose is to make them put on mass. Thus, at this level, increasing the energy, while barely increasing the speed, increases the hitting power by increasing the mass.

[1] The electron volt (eV) is the energy an electron acquires if accelerated in the field of 1V. In a television tube the glow is produced by electrons striking the screen with about 10,000 eV. One BeV is 1,000,000,000 eV, and 6 BeV is approximately equal to one watt second. In other words, each proton, minute as it is (there are 21 million million million million proton masses in one ounce), carries as much energy as a small light bulb (5 watts) emits in heat and light in one second.

Having seen that the energy of the motion of these particles shows itself to have mass, two questions come to mind:

(1) Do all forms of energy (such as light and other radiation, nuclear energies, etc.) have mass or is this property confined to the energy of motion?

(2) Given that some of the mass of these particles represents energy, does the rest mass (the mass they have when standing still) also represent some form of energy?

As for (1), clearly all forms of energy have mass. For the most characteristic property of energy is its interchangeability (consider, for example, the chain: chemical energy of coal–heat energy of steam in a power station–electrical energy in the wires–energy of motion of an electric train). If a change from one form of energy to another changed the mass it would play havoc with the laws of conservation of momentum, which is an explicit inference from experience and takes no note of internal changes.

Imagine a spaceship, with its engines off, traveling at constant speed. The people in the spaceship use the energy stored, say, in its electric batteries, at one time to cook their food, at another to drive their washing machine. By the conservation law, these internal transactions cannot change the momentum of the system. Thus if the spaceship is at rest relative to one inertial observer before these operations, then it must be at rest relative to him afterward. Relative to another inertial observer it has therefore not changed speed. Since its momentum is also the same, the mass must have stayed constant, and hence the energy when in the battery must have had the same mass as when it was energy of motion in the washing machine (or energy of heat in the cooker, etc.).

EINSTEIN'S EQUATION

Hence every form of energy has mass, given, as in the case of energy of motion, by Einstein's famous relation

$$E = mc^2$$
$$(\text{Energy}) = (\text{Mass}) \times (\text{Velocity of light})^2$$

In some of his own early work on Relativity Einstein used another method to establish that energy has mass. First consider how we become aware of mass rather than weight (which is due to our particular local condition of living in a gravitational field). Essentially our awareness of mass is due to its relation to force, and in Chapter II we saw how this relation led to considerations of momentum. The concept of momentum is so important because it applies to a system as a whole, regardless of what may go on inside the system (the baby and the baby carriage of Chapter II). In particular, if no *external* force acts on a system, then its momentum cannot change regardless of what goes on inside the system. Since the momentum governs the motion of the center of mass of the system, then if this center is initially at rest, it will remain at rest whatever may go on inside the system, provided no force acts on the system from outside. All this is just a precise statement of the saying that one cannot pull oneself up by one's own bootstraps.

As a special example, consider a long box lying on a smooth horizontal table (see Fig. 32). If nobody pushes the box, its center of mass, if initially at rest, will remain at rest whatever may go on inside the box. But this does not necessarily mean that the outside of the box will always remain at rest. If masses *inside* the box shift around, the location of the center of mass relative

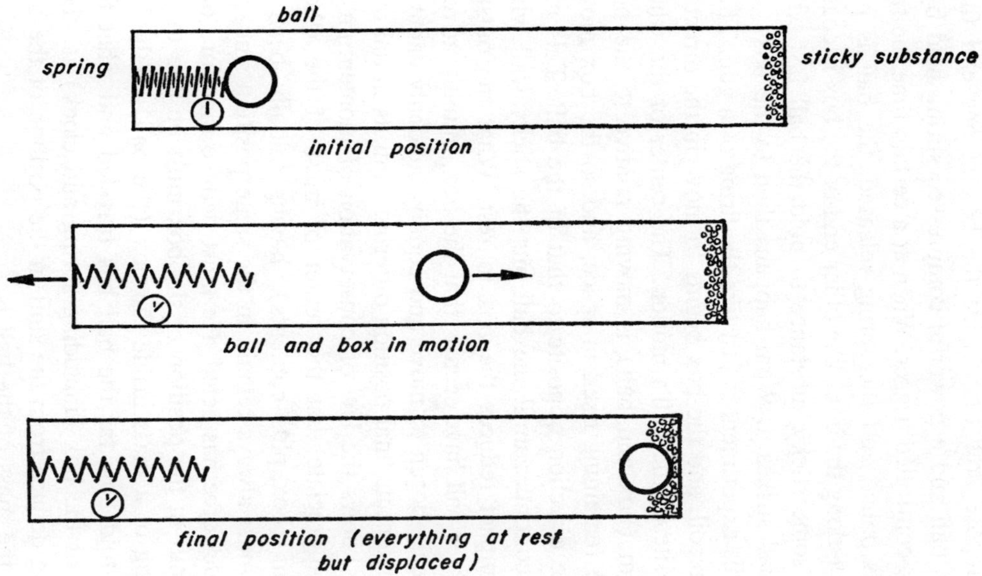
ball
spring
sticky substance
initial position
ball and box in motion
final position (everything at rest
but displaced)

FIG. 32.

to the box may change and then, with the center of mass remaining in position, the box will move.

Suppose now that inside the box at one end there are a ball and a powerful compressed spring held by a clock-controlled trigger. When at a certain time the trigger is pulled and the spring released, the ball is propelled along the box. The other end of the box is coated with some sticky substance to hold the ball fast when the ball strikes it. What happens then to the box?

This experiment is just like the firing of a gun. There is a recoil, and the box begins to move in the direction opposite to the ball's motion. The center of the whole system (box and ball) is moving relative to the box, while remaining fixed in space, and so the box moves in the direction opposite to that of the ball. This motion continues until the ball hits the sticky wall when the impact reduces the box to rest. What the outside observer will have seen is that the box, initially at rest, suddenly began to move and equally suddenly came to rest again, its final position differing from its initial one. If he knows the law of conservation of momentum he will appreciate that the center of mass of the whole system (box plus contents), having initially been at rest, must always remain in the same position since no outside force has acted. He must deduce therefore that the shift in the position of the box must be due to the shifting of a mass inside the box (the ball). If he has been told how far the ball has traveled and what the mass of the box (including spring and clock) is, then he can deduce from the shift in the position of the box what the mass of the ball is.

This example is a direct application of Newtonian dynamics, and Newton himself could easily have carried it out. Einstein's new insight came when he replaced the ball in our example with a flash of light.

The important property of light required for Einstein's demonstration is that it exerts a pressure. If light hits a black surface (so that it is absorbed) it gives the surface a push; if it hits a mirror (so that it bounces off) it gives twice as much push. For any reasonable intensity of light this pressure is quite small, but the existence of the pressure follows directly from Maxwell's theory of light (which came 40 years before Relativity) and can be demonstrated if enough care is taken. An apparatus in which light pushes a little paddle wheel round is a favorite model in science museums, and opticians occasionally display one in the window.

Suppose we have the same box but with the clock now operating a switch which connects a battery to a flashbulb emitting a short intense flash of light (Fig. 33). All the walls of the box are shiny and reflect light, except the wall at the far end from the bulb, which is black. When the switch is closed the bulb emits light in all directions. With the bulb close to one end, half the light bounces off this end and exerts a pressure on it, which sets the box in motion. When the flash hits the black end a little later (for light takes some time even to travel along a box!) all the light now exerts a pressure which reduces the box to rest. To the outside observer, therefore, the situation is in principle identical with that of the ball. The box, initially at rest, suddenly starts to move and then comes to rest again in a different position. The observer must therefore deduce that *mass has been transferred* from the bulb end to the black end, and he can calculate the quantity of mass from the displacement of the box. Maxwell's theory of light shows that the pressure of light on a black surface equals its intensity divided by the speed of light. Combining this relationship with the travel time of light and the duration of the flash, Einstein found that the mass

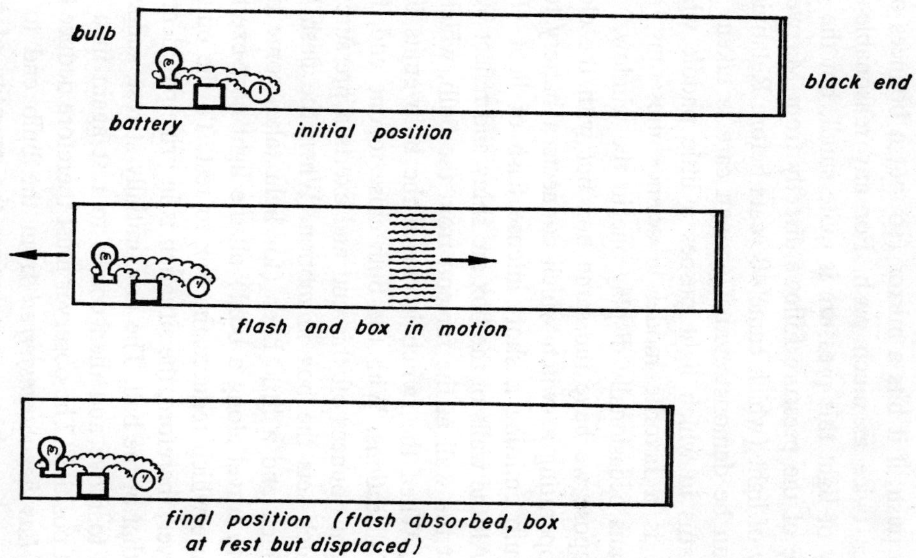
bulb
black end
battery
initial position
flash and box in motion
final position (flash absorbed, box
at rest but displaced)

FIG. 33.

transferred equals the energy of the flash divided by the square of the speed of light.

Unquestionably energy has been transferred from the bulb end of the box (where originally the energy was stored in the battery) to the black end, which has been heated by absorbing the light. What Einstein's ideal experiment thus shows is that this transfer of energy E is accompanied by a transfer of mass m, these two quantities being related by

$$E = mc^2$$

Thus the energy of light, just like the energy of motion, has mass. Again, starting from this knowledge, we can deduce, as before, that *all* energy must have mass in accordance with this relation.

To come now to the question whether rest mass, too, represents energy, we have to turn to nuclear physics. All nuclei consist of protons and neutrons. The mass of a composite nucleus is less (by almost as much as 1 per-cent) than the sum of the masses of the protons and neutrons of which it is built. This difference is accounted for by the energy released (and radiated away) when the protons and neutrons fuse to form the composite nucleus. Here is the clue to nuclear energies (atom bombs, nuclear power stations), and proves the complete equivalence of mass and energy. Thus Einstein's theory has not only unified optics and dynamics, has not only clarified the meaning of time and space, but also has unified the concepts of mass and energy.

Theory and Observation

This concludes our brief survey of the Theory of Relativity. I hope to have shown that this theory, at one stage considered so mysterious, is in fact the most obvious and clear-cut extension of ordinary ideas to the realm of high velocities. What is unfamiliar in it is unfamiliar only because high velocities are unfamiliar. No other way could have made the world of high velocities so simple and so intelligible. But the chief purpose of a scientific theory is not just to be simple and intelligible, is not just to enrich experience and unify observations; it must also fit the facts, and here, we may say, there is perhaps no other part of physics that has been checked and tested and cross-checked quite as much as the Theory of Relativity. Naturally, not every one of the ideal experiments that, for the sake of simplicity, we have been considering has itself been tested, but there are so many points of contact between the theory and the observations that within the context of what we have set out to do there is really no doubt that this theory describes correctly this particular range of experiences.

Wherever high speeds occur, whether in particle accelerators or in optics, everywhere the Theory of Relativity has stood the test of observation perfectly. It has added to our understanding of nature a framework into which we believe that all our physical theories must fit. If we have been able to show that the theory is not difficult and mysterious, this book will have served its purpose.

INDEX

A CATALOGUE OF SELECTED DOVER BOOKS IN ALL FIELDS OF INTEREST

A CATALOGUE OF SELECTED DOVER BOOKS IN ALL FIELDS OF INTEREST

DRAWINGS OF WILLIAM BLAKE, William Blake. 92 plates from Book of Job, *Divine Comedy, Paradise Lost,* visionary heads, mythological figures, Laocoon, etc. Selection, introduction, commentary by Sir Geoffrey Keynes. 178pp. 8⅛ x 11. 22303-5 Pa. $5.00

ENGRAVINGS OF HOGARTH, William Hogarth. 101 of Hogarth's greatest works: *Rake's Progress, Harlot's Progress, Illustrations for Hudibras, Before and After, Beer Street and Gin Lane,* many more. Full commentary. 256pp. 11 x 13¾. 22479-1 Pa. $12.95

DAUMIER: 120 GREAT LITHOGRAPHS, Honore Daumier. Wide-ranging collection of lithographs by the greatest caricaturist of the 19th century. Concentrates on eternally popular series on lawyers, on married life, on liberated women, etc. Selection, introduction, and notes on plates by Charles F. Ramus. Total of 158pp. 9⅜ x 12¼. 23512-2 Pa. $6.00

DRAWINGS OF MUCHA, Alphonse Maria Mucha. Work reveals draftsman of highest caliber: studies for famous posters and paintings, renderings for book illustrations and ads, etc. 70 works, 9 in color; including 6 items not drawings. Introduction. List of illustrations. 72pp. 9⅜ x 12¼. (Available in U.S. only) 23672-2 Pa. $4.50

GIOVANNI BATTISTA PIRANESI: DRAWINGS IN THE PIERPONT MORGAN LIBRARY, Giovanni Battista Piranesi. For first time ever all of Morgan Library's collection, world's largest. 167 illustrations of rare Piranesi drawings—archeological, architectural, decorative and visionary. Essay, detailed list of drawings, chronology, captions. Edited by Felice Stampfle. 144pp. 9⅜ x 12¼. 23714-1 Pa. $7.50

NEW YORK ETCHINGS (1905-1949), John Sloan. All of important American artist's N.Y. life etchings. 67 works include some of his best art; also lively historical record—Greenwich Village, tenement scenes. Edited by Sloan's widow. Introduction and captions. 79pp. 8⅜ x 11¼. 23651-X Pa. $5.00

CHINESE PAINTING AND CALLIGRAPHY: A PICTORIAL SURVEY, Wan-go Weng. 69 fine examples from John M. Crawford's matchless private collection: landscapes, birds, flowers, human figures, etc., plus calligraphy. Every basic form included: hanging scrolls, handscrolls, album leaves, fans, etc. 109 illustrations. Introduction. Captions. 192pp. 8⅞ x 11¾. 23707-9 Pa. $7.95

DRAWINGS OF REMBRANDT, edited by Seymour Slive. Updated Lippmann, Hofstede de Groot edition, with definitive scholarly apparatus. All portraits, biblical sketches, landscapes, nudes, Oriental figures, classical studies, together with selection of work by followers. 550 illustrations. Total of 630pp. 9⅛ x 12¼. 21485-0, 21486-9 Pa., Two-vol. set $17.90

THE DISASTERS OF WAR, Francisco Goya. 83 etchings record horrors of Napoleonic wars in Spain and war in general. Reprint of 1st edition, plus 3 additional plates. Introduction by Philip Hofer. 97pp. 9⅜ x 8¼. 21872-4 Pa. $4.50

THE EARLY WORK OF AUBREY BEARDSLEY, Aubrey Beardsley. 157 plates, 2 in color: *Manon Lescaut, Madame Bovary, Morte Darthur, Salome,* other. Introduction by H. Marillier. 182pp. 8⅛ x 11. 21816-3 Pa. **$6.50**

THE LATER WORK OF AUBREY BEARDSLEY, Aubrey Beardsley. Exotic masterpieces of full maturity: *Venus and Tannhauser, Lysistrata, Rape of the Lock, Volpone,* Savoy material, etc. 174 plates, 2 in color. 186pp. 8⅛ x 11. 21817-1 Pa. $5.95

THOMAS NAST'S CHRISTMAS DRAWINGS, Thomas Nast. Almost all Christmas drawings by creator of image of Santa Claus as we know it, and one of America's foremost illustrators and political cartoonists. 66 illustrations. 3 illustrations in color on covers. 96pp. 8⅜ x 11¼. 23660-9 Pa. $3.50

THE DORÉ ILLUSTRATIONS FOR DANTE'S DIVINE COMEDY, Gustave Doré. All 135 plates from Inferno, Purgatory, Paradise; fantastic tortures, infernal landscapes, celestial wonders. Each plate with appropriate (translated) verses. 141pp. 9 x 12. 23231-X Pa. **$5.00**

DORÉ'S ILLUSTRATIONS FOR RABELAIS, Gustave Doré. 252 striking illustrations of *Gargantua and Pantagruel* books by foremost 19th-century illustrator. Including 60 plates, 192 delightful smaller illustrations. 153pp. **9 x 12.** **23656-0 Pa. $6.00**

LONDON: A PILGRIMAGE, Gustave Doré, Blanchard Jerrold. Squalor, riches, misery, beauty of mid-Victorian metropolis; 55 wonderful plates, 125 other illustrations, full social, cultural text by Jerrold. 191pp. of text. 9⅜ x 12¼. 22306-X Pa. **$7.00**

THE RIME OF THE ANCIENT MARINER, Gustave Doré, S. T. Coleridge. Dore's finest work, 34 plates capture moods, subtleties of poem. Full text. Introduction by Millicent Rose. 77pp. 9¼ x 12. 22305-1 Pa. $4.50

THE DORE BIBLE ILLUSTRATIONS, Gustave Doré. All wonderful, detailed plates: Adam and Eve, Flood, Babylon, Life of Jesus, etc. Brief King James text with each plate. Introduction by Millicent Rose. 241 plates. 241pp. 9 x 12. 23004-X Pa. $6.95

THE COMPLETE ENGRAVINGS, ETCHINGS AND DRYPOINTS OF ALBRECHT DURER. "Knight, Death and Devil"; "Melencolia," and more—all Dürer's known works in all three media, including 6 works formerly attributed to him. 120 plates. 235pp. 8⅜ x 11¼. 22851-7 Pa. **$7.50**

MECHANICK EXERCISES ON THE WHOLE ART OF PRINTING, Joseph Moxon. First complete book (1683-4) ever written about typography, a compendium of everything known about printing at the latter part of 17th century. Reprint of 2nd (1962) Oxford Univ. Press edition. 74 illustrations. Total of 550pp. 6⅛ x 9¼. 23617-X Pa. $7.95

CATALOGUE OF DOVER BOOKS

THE COMPLETE WOODCUTS OF ALBRECHT DURER, edited by Dr. W. Kurth. 346 in all: "Old Testament," "St. Jerome," "Passion," "Life of Virgin," Apocalypse," many others. Introduction by Campbell Dodgson. 285pp. 8½ x 12¼. 21097-9 Pa. $7.50

DRAWINGS OF ALBRECHT DURER, edited by Heinrich Wolfflin. 81 plates show development from youth to full style. Many favorites; many new. Introduction by Alfred Werner. 96pp. 8⅛ x 11. 22352-3 Pa. $6.00

THE HUMAN FIGURE, Albrecht Dürer. Experiments in various techniques—stereometric, progressive proportional, and others. Also life studies that rank among finest ever done. Complete reprinting of *Dresden Sketchbook*. 170 plates. 355pp. 8⅜ x 11¼. 21042-1 Pa. $7.95

OF THE JUST SHAPING OF LETTERS, Albrecht Dürer. Renaissance artist explains design of Roman majuscules by geometry, also Gothic lower and capitals. Grolier Club edition. 43pp. 7⅞ x 10¾ 21306-4 Pa. $3.00

TEN BOOKS ON ARCHITECTURE, Vitruvius. The most important book ever written on architecture. Early Roman aesthetics, technology, classical orders, site selection, all other aspects. Stands behind everything since. Morgan translation. 331pp. 5⅜ x 8½. 20645-9 Pa. $5.00

THE FOUR BOOKS OF ARCHITECTURE, Andrea Palladio. 16th-century classic responsible for Palladian movement and style. Covers classical architectural remains, Renaissance revivals, classical orders, etc. 1738 Ware English edition. Introduction by A. Placzek. 216 plates. 110pp. of text. 9½ x 12¾. 21308-0 Pa. $10.00

HORIZONS, Norman Bel Geddes. Great industrialist stage designer, "father of streamlining," on application of aesthetics to transportation, amusement, architecture, etc. 1932 prophetic account; function, theory, specific projects. 222 illustrations. 312pp. 7⅞ x 10¾. 23514-9 Pa. $6.95

FRANK LLOYD WRIGHT'S FALLINGWATER, Donald Hoffmann. Full, illustrated story of conception and building of Wright's masterwork at Bear Run, Pa. 100 photographs of site, construction, and details of completed structure. 112pp. 9¼ x 10. 23671-4 Pa. $5.95

THE ELEMENTS OF DRAWING, John Ruskin. Timeless classic by great Viltorian; starts with basic ideas, works through more difficult. Many practical exercises. 48 illustrations. Introduction by Lawrence Campbell. 228pp. 5⅜ x 8½. 22730-8 Pa. $3.75

GIST OF ART, John Sloan. Greatest modern American teacher, Art Students League, offers innumerable hints, instructions, guided comments to help you in painting. Not a formal course. 46 illustrations. Introduction by Helen Sloan. 200pp. 5⅜ x 8½. 23435-5 Pa. $4.00

ART FORMS IN NATURE, Ernst Haeckel. Multitude of strangely beautiful natural forms: Radiolaria, Foraminifera, jellyfishes, fungi, turtles, bats, etc. All 100 plates of the 19th-century evolutionist's *Kunstformen der Natur* (1904). 100pp. 9⅜ x 12¼. 22987-4 Pa. $5.00

CHILDREN: A PICTORIAL ARCHIVE FROM NINETEENTH-CENTURY SOURCES, edited by Carol Belanger Grafton. 242 rare, copyright-free wood engravings for artists and designers. Widest such selection available. All illustrations in line. 119pp. 8⅜ x 11¼. 23694-3 Pa. $4.00

WOMEN: A PICTORIAL ARCHIVE FROM NINETEENTH-CENTURY SOURCES, edited by Jim Harter. 391 copyright-free wood engravings for artists and designers selected from rare periodicals. Most extensive such collection available. All illustrations in line. 128pp. 9 x 12. 23703-6 Pa. $4.95

ARABIC ART IN COLOR, Prisse d'Avennes. From the greatest ornamentalists of all time—50 plates in color, rarely seen outside the Near East, rich in suggestion and stimulus. Includes 4 plates on covers. 46pp. 9⅜ x 12¼. 23658-7 Pa. $6.00

AUTHENTIC ALGERIAN CARPET DESIGNS AND MOTIFS, edited by June Beveridge. Algerian carpets are world famous. Dozens of geometrical motifs are charted on grids, color-coded, for weavers, needleworkers, craftsmen, designers. 53 illustrations plus 4 in color. 48pp. 8¼ x 11. (Available in U.S. only) 23650-1 Pa. $1.75

DICTIONARY OF AMERICAN PORTRAITS, edited by Hayward and Blanche Cirker. 4000 important Americans, earliest times to 1905, mostly in clear line. Politicians, writers, soldiers, scientists, inventors, industrialists, Indians, Blacks, women, outlaws, etc. Identificatory information. 756pp. 9¼ x 12¾. 21823-6 Clothbd. $65.00

HOW THE OTHER HALF LIVES, Jacob A. Riis. Journalistic record of filth, degradation, upward drive in New York immigrant slums, shops, around 1900. New edition includes 100 original Riis photos, monuments of early photography. 233pp. 10 x 7⅞. 22012-5 Pa. $7.00

NEW YORK IN THE THIRTIES, Berenice Abbott. Noted photographer's fascinating study of city shows new buildings that have become famous and old sights that have disappeared forever. Insightful commentary. 97 photographs. 97pp. 11⅜ x 10. 22967-X Pa. $6.00

MEN AT WORK, Lewis W. Hine. Famous photographic studies of construction workers, railroad men, factory workers and coal miners. New supplement of 18 photos on Empire State building construction. New introduction by Jonathan L. Doherty. Total of 69 photos. 63pp. 8 x 10¾. 23475-4 Pa. $4.00

THE DEPRESSION YEARS AS PHOTOGRAPHED BY ARTHUR ROTHSTEIN, Arthur Rothstein. First collection devoted entirely to the work of outstanding 1930s photographer: famous dust storm photo, ragged children, unemployed, etc. 120 photographs. Captions. 119pp. 9¼ x 10¾.
23590-4 Pa. **$5.95**

CAMERA WORK: A PICTORIAL GUIDE, Alfred Stieglitz. All 559 illustrations and plates from the most important periodical in the history of art photography, *Camera Work* (1903-17). Presented four to a page, reduced in size but still clear, in strict chronological order, with complete captions. Three indexes. Glossary. Bibliography. 176pp. 8⅜ x 11¼.
23591-2 Pa. $6.95

ALVIN LANGDON COBURN, PHOTOGRAPHER, Alvin L. Coburn. Revealing autobiography by one of greatest photographers of 20th century gives insider's version of Photo-Secession, plus comments on his own work. 77 photographs by Coburn. Edited by Helmut and Alison Gernsheim. 160pp. 8⅛ x 11. 23685-4 Pa. $6.00

NEW YORK IN THE FORTIES, Andreas Feininger. 162 brilliant photographs by the well-known photographer, formerly with *Life* magazine, show commuters, shoppers, Times Square at night, Harlem nightclub, Lower East Side, etc. Introduction and full captions by John von Hartz. 181pp. 9¼ x 10¾. 23585-8 Pa. $6.95

GREAT NEWS PHOTOS AND THE STORIES BEHIND THEM, John Faber. Dramatic volume of 140 great news photos, 1855 through 1976, and revealing stories behind them, with both historical and technical information. Hindenburg disaster, shooting of Oswald, nomination of Jimmy Carter, etc. 160pp. 8¼ x 11. 23667-6 Pa. **$6.00**

THE ART OF THE CINEMATOGRAPHER, Leonard Maltin. Survey of American cinematography history and anecdotal interviews with 5 masters—Arthur Miller, Hal Mohr, Hal Rosson, Lucien Ballard, and Conrad Hall. Very large selection of behind-the-scenes production photos. 105 photographs. Filmographies. Index. Originally *Behind the Camera.* 144pp. 8¼ x 11. 23686-2 Pa. $5.00

DESIGNS FOR THE THREE-CORNERED HAT (LE TRICORNE), Pablo Picasso. 32 fabulously rare drawings—including 31 color illustrations of costumes and accessories—for 1919 production of famous ballet. Edited by Parmenia Migel, who has written new introduction. 48pp. 9⅜ x 12¼. (Available in U.S. only) 23709-5 Pa. $5.00

NOTES OF A FILM DIRECTOR, Sergei Eisenstein. Greatest Russian filmmaker explains montage, making of *Alexander Nevsky,* aesthetics; comments on self, associates, great rivals (Chaplin), similar material. 78 illustrations. 240pp. 5⅜ x 8½. 22392-2 Pa. $7.00

A MAYA GRAMMAR, Alfred M. Tozzer. Practical, useful English-language grammar by the Harvard anthropologist who was one of the three greatest American scholars in the area of Maya culture. Phonetics, grammatical processes, syntax, more. 301pp. 5⅜ x 8½. 23465-7 Pa. $4.00

THE JOURNAL OF HENRY D. THOREAU, edited by Bradford Torrey, F. H. Allen. Complete reprinting of 14 volumes, 1837-61, over two million words; the sourcebooks for *Walden*, etc. Definitive. All original sketches, plus 75 photographs. Introduction by Walter Harding. Total of 1804pp. 8½ x 12¼. 20312-3, 20313-1 Clothbd., Two-vol. set $80.00

CLASSIC GHOST STORIES, Charles Dickens and others. 18 wonderful stories you've wanted to reread: "The Monkey's Paw," "The House and the Brain," "The Upper Berth," "The Signalman," "Dracula's Guest," "The Tapestried Chamber," etc. Dickens, Scott, Mary Shelley, Stoker, etc. 330pp. 5⅜ x 8½. 20735-8 Pa. $4.50

SEVEN SCIENCE FICTION NOVELS, H. G. Wells. Full novels. *First Men in the Moon, Island of Dr. Moreau, War of the Worlds, Food of the Gods, Invisible Man, Time Machine, In the Days of the Comet.* A basic science-fiction library. 1015pp. 5⅜ x 8½. (Available in U.S. only) 20264-X Clothbd. $15.00

ARMADALE, Wilkie Collins. Third great mystery novel by the author of *The Woman in White* and *The Moonstone.* Ingeniously plotted narrative shows an exceptional command of character, incident and mood. Original magazine version with 40 illustrations. 597pp. 5⅜ x 8½. 23429-0 Pa. $7.95

FLATLAND, E. A. Abbott. Science-fiction classic explores life of 2-D being in 3-D world. Read also as introduction to thought about hyperspace. Introduction by Banesh Hoffmann. 16 illustrations. 103pp. 5⅜ x 8½. 20001-9 Pa. $2.75

AYESHA: THE RETURN OF "SHE," H. Rider Haggard. Virtuoso sequel featuring the great mythic creation, Ayesha, in an adventure that is fully as good as the first book, *She.* Original magazine version, with 47 original illustrations by Maurice Greiffenhagen. 189pp. 6½ x 9¼. 23649-8 Pa. $3.50

ORIENTAL RUGS, ANTIQUE AND MODERN, Walter A. Hawley. Persia, Turkey, Caucasus, Central Asia, China, other traditions. Best general survey of all aspects: styles and periods, manufacture, uses, symbols and their interpretation, and identification. 96 illustrations, 11 in color. 320pp. 6⅛ x 9¼. 22366-3 Pa. $6.95

CHINESE POTTERY AND PORCELAIN, R. L. Hobson. Detailed descriptions and analyses by former Keeper of the Department of Oriental Antiquities and Ethnography at the British Museum. Covers hundreds of pieces from primitive times to 1915. Still the standard text for most periods. 136 plates, 40 in full color. Total of 750pp. 5⅜ x 8½. 23253-0 Pa. $10.00

UNCLE SILAS, J. Sheridan LeFanu. Victorian Gothic mystery novel, considered by many best of period, even better than Collins or Dickens. Wonderful psychological terror. Introduction by Frederick Shroyer. 436pp. 5⅜ x 8½. 21715-9 Pa. **$6.95**

JURGEN, James Branch Cabell. The great erotic fantasy of the 1920's that delighted thousands, shocked thousands more. Full final text, Lane edition with 13 plates by Frank Pape. 346pp. 5⅜ x 8½. 23507-6 Pa. $4.50

THE CLAVERINGS, Anthony Trollope. Major novel, chronicling aspects of British Victorian society, personalities. Reprint of Cornhill serialization, 16 plates by M. Edwards; first reprint of full text. Introduction by Norman Donaldson. 412pp. 5⅜ x 8½. 23464-9 Pa. $5.00

KEPT IN THE DARK, Anthony Trollope. Unusual short novel about Victorian morality and abnormal psychology by the great English author. Probably the first American publication. Frontispiece by Sir John Millais. 92pp. 6½ x 9¼. 23609-9 Pa. $2.50

RALPH THE HEIR, Anthony Trollope. Forgotten tale of illegitimacy, inheritance. Master novel of Trollope's later years. Victorian country estates, clubs, Parliament, fox hunting, world of fully realized characters. Reprint of 1871 edition. 12 illustrations by F. A. Faser. 434pp. of text. 5⅜ x 8½. 23642-0 Pa. $6.50

YEKL and THE IMPORTED BRIDEGROOM AND OTHER STORIES OF THE NEW YORK GHETTO, Abraham Cahan. Film *Hester Street* based on *Yekl* (1896). Novel, other stories among first about Jewish immigrants of N.Y.'s East Side. Highly praised by W. D. Howells—Cahan "a new star of realism." New introduction by Bernard G. Richards. 240pp. 5⅜ x 8½. 22427-9 Pa. $3.50

THE HIGH PLACE, James Branch Cabell. Great fantasy writer's enchanting comedy of disenchantment set in 18th-century France. Considered by some critics to be even better than his famous *Jurgen*. 10 illustrations and numerous vignettes by noted fantasy artist Frank C. Pape. 320pp. 5⅜ x 8½. 23670-6 Pa. $4.00

ALICE'S ADVENTURES UNDER GROUND, Lewis Carroll. Facsimile of ms. Carroll gave Alice Liddell in 1864. Different in many ways from final Alice. Handlettered, illustrated by Carroll. Introduction by Martin Gardner. 128pp. 5⅜ x 8½. 21482-6 Pa. $2.50

FAVORITE ANDREW LANG FAIRY TALE BOOKS IN MANY COLORS, Andrew Lang. The four Lang favorites in a boxed set—the complete *Red, Green, Yellow* and *Blue* Fairy Books. 164 stories; 439 illustrations by Lancelot Speed, Henry Ford and G. P. Jacomb Hood. Total of about 1500pp. 5⅜ x 8½. 23407-X Boxed set, Pa. $16.95

HOUSEHOLD STORIES BY THE BROTHERS GRIMM. All the great Grimm stories: "Rumpelstiltskin," "Snow White," "Hansel and Gretel," etc., with 114 illustrations by Walter Crane. 269pp. 5⅜ x 8½.
21080-4 Pa. $3.50

SLEEPING BEAUTY, illustrated by Arthur Rackham. Perhaps the fullest, most delightful version ever, told by C. S. Evans. Rackham's best work. 49 illustrations. 110pp. 7⅞ x 10¾. 22756-1 Pa. $2.95

AMERICAN FAIRY TALES, L. Frank Baum. Young cowboy lassoes Father Time; dummy in Mr. Floman's department store window comes to life; and 10 other fairy tales. 41 illustrations by N. P. Hall, Harry Kennedy, Ike Morgan, and Ralph Gardner. 209pp. 5⅜ x 8½. 23643-9 Pa. $3.00

THE WONDERFUL WIZARD OF OZ, L. Frank Baum. Facsimile in full color of America's finest children's classic. Introduction by Martin Gardner. 143 illustrations by W. W. Denslow. 267pp. 5⅜ x 8½.
20691-2 Pa. $4.50

THE TALE OF PETER RABBIT, Beatrix Potter. The inimitable Peter's terrifying adventure in Mr. McGregor's garden, with all 27 wonderful, full-color Potter illustrations. 55pp. 4¼ x 5½. (Available in U.S. only)
22827-4 Pa. $1.50

THE STORY OF KING ARTHUR AND HIS KNIGHTS, Howard Pyle. Finest children's version of life of King Arthur. 48 illustrations by Pyle. 131pp. 6⅛ x 9¼. 21445-1 Pa. $5.95

CARUSO'S CARICATURES, Enrico Caruso. Great tenor's remarkable caricatures of self, fellow musicians, composers, others. Toscanini, Puccini, Farrar, etc. Impish, cutting, insightful. 473 illustrations. Preface by M. Sisca. 217pp. 8⅜ x 11¼. 23528-9 Pa. $6.95

PERSONAL NARRATIVE OF A PILGRIMAGE TO ALMADINAH AND MECCAH, Richard Burton. Great travel classic by remarkably colorful personality. Burton, disguised as a Moroccan, visited sacred shrines of Islam, narrowly escaping death. Wonderful observations of Islamic life, customs, personalities. 47 illustrations. Total of 959pp. 5⅜ x 8½.
21217-3, 21218-1 Pa., Two-vol. set $14.00

INCIDENTS OF TRAVEL IN YUCATAN, John L. Stephens. Classic (1843) exploration of jungles of Yucatan, looking for evidences of Maya civilization. Travel adventures, Mexican and Indian culture, etc. Total of 669pp. 5⅜ x 8½. 20926-1, 20927-X Pa., Two-vol. set $7.90

AMERICAN LITERARY AUTOGRAPHS FROM WASHINGTON IRVING TO HENRY JAMES, Herbert Cahoon, et al. Letters, poems, manuscripts of Hawthorne, Thoreau, Twain, Alcott, Whitman, 67 other prominent American authors. Reproductions, full transcripts and commentary. Plus checklist of all American Literary Autographs in The Pierpont Morgan Library. Printed on exceptionally high-quality paper. 136 illustrations. 212pp. 9⅛ x 12¼. 23548-3 Pa. $12.50

AN AUTOBIOGRAPHY, Margaret Sanger. Exciting personal account of hard-fought battle for woman's right to birth control, against prejudice, church, law. Foremost feminist document. 504pp. 5⅜ x 8½.
20470-7 Pa. $7.50

MY BONDAGE AND MY FREEDOM, Frederick Douglass. Born as a slave, Douglass became outspoken force in antislavery movement. The best of Douglass's autobiographies. Graphic description of slave life. Introduction by P. Foner. 464pp. 5⅜ x 8½. 22457-0 Pa. $6.50

LIVING MY LIFE, Emma Goldman. Candid, no holds barred account by foremost American anarchist: her own life, anarchist movement, famous contemporaries, ideas and their impact. Struggles and confrontations in America, plus deportation to U.S.S.R. Shocking inside account of persecution of anarchists under Lenin. 13 plates. Total of 944pp. 5⅜ x 8½.
22543-7, 22544-5 Pa., Two-vol. set $12.00

LETTERS AND NOTES ON THE MANNERS, CUSTOMS AND CONDITIONS OF THE NORTH AMERICAN INDIANS, George Catlin. Classic account of life among Plains Indians: ceremonies, hunt, warfare, etc. Dover edition reproduces for first time all original paintings. 312 plates. 572pp. of text. 6⅛ x 9¼. 22118-0, 22119-9 Pa.. Two-vol. set $12.00

THE MAYA AND THEIR NEIGHBORS, edited by Clarence L. Hay, others. Synoptic view of Maya civilization in broadest sense, together with Northern, Southern neighbors. Integrates much background, valuable detail not elsewhere. Prepared by greatest scholars: Kroeber, Morley, Thompson, Spinden, Vaillant, many others. Sometimes called Tozzer Memorial Volume. 60 illustrations, linguistic map. 634pp. 5⅜ x 8½.
23510-6 Pa. $10.00

HANDBOOK OF THE INDIANS OF CALIFORNIA, A. L. Kroeber. Foremost American anthropologist offers complete ethnographic study of each group. Monumental classic. 459 illustrations, maps. 995pp. 5⅜ x 8½.
23368-5 Pa. $13.00

SHAKTI AND SHAKTA, Arthur Avalon. First book to give clear, cohesive analysis of Shakta doctrine, Shakta ritual and Kundalini Shakti (yoga). Important work by one of world's foremost students of Shaktic and Tantric thought. 732pp. 5⅜ x 8½. (Available in U.S. only)
23645-5 Pa. $7.95

AN INTRODUCTION TO THE STUDY OF THE MAYA HIEROGLYPHS, Syvanus Griswold Morley. Classic study by one of the truly great figures in hieroglyph research. Still the best introduction for the student for reading Maya hieroglyphs. New introduction by J. Eric S. Thompson. 117 illustrations. 284pp. 5⅜ x 8½. 23108-9 Pa. $4.00

A STUDY OF MAYA ART, Herbert J. Spinden. Landmark classic interprets Maya symbolism, estimates styles, covers ceramics, architecture, murals, stone carvings as artforms. Still a basic book in area. New introduction by J. Eric Thompson. Over 750 illustrations. 341pp. 8⅜ x 11¼.
21235-1 Pa. $6.95

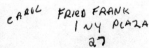

CAROL FRIED FRANK
1 NY PLAZA
27

PA WTRADE
JERRY 1 WT CTR
55